RALPH

CLARENCE RALPH FITZ
MARDELLE MARIE FITZ MEYER

ISBN: 978-1-955622-50-9 (hardcover)
978-1-962402-29-3 (paperback)

Published by
Fideli Publishing, Inc.

www.FideliPublishing.com

Dedicated to the memory of
Edith Aleta Hill Fitz
and
Ralph Glenn Fitz
Our parents

Picture taken in Bremerton in 1917.

Ralph
Glenn
Fitz

Navy whites.

TABLE OF CONTENTS

PREFACE I

It was several years ago, probably after finding and reading Dad Ralph's journals or diaries and all of the letters that he saved that recorded his Montana days and his Navy days, that I promised myself that I would write a book about his homesteading and naval career. I worked on that first sentence, "Ralph had plenty of time to think as he walked down the dusty road from his new homestead on his way to the train station and the train that would take him to the induction point into the United States Navy." As you will see, the book doesn't start with that sentence.

For the next probably about ten years, I got involved in the research for and the writing of two books on Federal Indian Policy as it related to my lake home on Mille Lacs Lake in Minnesota. Those two books, "...And the Mille Lacs who have no reservation..." and The Pendulum, from Indian Removal to Buying Mille Lacs gave me the experience needed to write this book as well as the confidence of a publisher who trusts me to write a book worth publishing.

Then two or so years ago when visiting my sister Mardelle and my brother-in-law Roger, I became aware of the extensive information my sister Mardelle had compiled on our family history, and so the seed was planted and this book is the cooperative effort of my sister Mardelle Marie Fitz Meyer and myself Clarence Ralph Fitz.

I, Clarence, did most of the rough draft writing, and while Mardelle's input has been invaluable, I must accept the blame for any errors that you find in either content or spelling or grammar, and for those errors I am regretful.

Both Mardelle and I want to thank the several ladies at the county courthouses of Greene County, Iowa, Calhoun County, Iowa, Franklin County, Iowa and Phillips County, Montana. While one county was reluctant, we finally got everything we were looking for to put the story together that we have written. We also want to thank the lady at the Malta, Montana Museum for her help with the plat maps of Phillips County where Ralph's Montana adventure and homestead were located. We also want to thank the folks at Transfer for Less in Mesa, Arizona and Mardelle's son Mark Meyer for digitalizing pictures and documents for us that were beyond the computer capability of our age level. Lastly, we want to thank those of our family who helped compile the information needed for the section on descendants of Ralph and Edith, our Mom and Dad.

Ralph was a self-made man who took the good examples set by his parents and grandparents, started with very little, cut his losses when necessary, endured the Great Depression without folding and turned a half section of Iowa farm land into a prosperous operation. In so doing he passed on his work ethic and sound judgement to his family and provided a good life for himself and his family. Serving his country when asked to do so may have saved him from getting mired in the Montana homesteading disaster. But maybe his keen observation after experiencing both Montana life and Iowa life helped him make the right decision. That is for you to ponder and decide after reading the book. This is his story.

Clarence Ralph Fitz

PREFACE II

When I received a letter from my brother Clarence saying he was thinking about writing a book about our father, Ralph Glenn Fitz and he was wondering if I wanted to help him, I didn't have to think twice, of course I would be happy to help in any way I could.

The search was on to find information about our father. I had pictures, letters, postcards and a few documents that I contributed. This was a new experience for me so at times I was frustrated, confused and emotional, but it has also been rewarding to have contributed to the book of my Father. I am sure our Father would be proud to know that his story has been published into a book.

Thank you, Clarence for having the talent to write a book about our Father.

Mardelle Marie Fitz Meyer

CHAPTER 1

JOLLY OLDE ENGLAND

I n her book, *Fitts Families (Fitts-Fitz-Fittz) A Genealogy*, Sylvia Fitts Getchell, a distant relative of the authors of this book, writes, "The earliest family member of whom we have some considerable knowledge was John Fitz, a lawyer and bencher of Lincoln's Inn." (1) pg 4 For the purposes of this book we will refer to this John Fitz as John Fitz the first as there were others that followed with the same name as well as to denote the generation regardless of name.

John Fitz the first was a lawyer and became a bencher at Lincoln's Inn located in London. Lincoln's Inn, a forerunner of what we now call a law school, was composed of students and lawyers and the benchers were the governing body. This was the 15th century and the legal profession was in its infancy at this time.

John Fitz the first was born near the town of Tavistock in western Devon. There are references to the Fitz name in the Devon area as early as 1220. John Fitz the first became prominent in the legal profession and was apparently rewarded well. "He settled in a manor house about a quarter of a mile outside of Tavistock where there was a

Fitzford Gatehouse

ford across the River Tavy. The manor house became known as Fitzford."
(1) pg6 Whether John Fitz the first only owned the manor house or whether
he was actually lord of the manor is unclear. But suffice it to say that a manor
was an economic unit with serfs and peasants under the jurisdiction of the
lord of the manor.

John Fitz the fourth was probably born at Fitzford about 1528, indicating that the manor house had remained in the family and he had inherited
it from his father John Fitz the third. John Fitz the fourth "was High Sheriff
of Devonshire in either [the] 23rd or 25th year of Queen Elizabeth's [the
First] reign." (1)pg13 He was a lawyer of great ability and "is said to have left
behind him a large volume of legal opinions in [a] manuscript entitled 'Fitz
his Reports." (1) pg13 In 1554 he was a representative in Parliament.

There is an interesting story involving John Fitz the fourth that has survived the test of time.

> An ancient story has it that John and his wife Mary were
> out riding and were led astray by pixies, to be lost in a bog
> on Dartmoor. They finally had given up finding their way,
> when they came upon a pure spring. After refreshing them-
> selves at the spring, the power of the pixies was dispelled
> and John Fitz and his wife rode safely off the moor. He had
> a stone erected above the spring in gratitude for their deliv-
> erance. The spring still called Fitz's well, is today within the
> grounds of Dartmoor Prison. (1) pg13

We will leave it to you the reader to ponder whether pixies actually
existed, but British folklore talks about them being particularly concentrated
in high moorland areas around Devon and Cornwall. There is evidence that
water from this spring was piped using lead pipes to the manor house at
Fitzford.

As evidenced by this story, John Fitz the fourth was apparently very
superstitious, as was common in sixteenth century England. When John Fitz
the fourth's son, his only son as it turned out, was about to be born, the stars
were telling him that this was an unlucky time to be born and he tried to
have his son's birth delayed. "He is reported to have asked the midwife to try
to delay the birth, if only for an hour, but this was not to be done." (1)pg16
John the fourth did not live long enough to see his son John the fifth become

a knight, probably the first knight in the Fitz family, and not long enough to see that the stars had been right. John Fitz the fourth died on January 2, 1589.

In an account apparently written by the chaplain to the Earl of Northumberland, the chaplain wrote:

> In 1599, 4 June, after a dinner and drinking party at Tavistock, a dispute arose between John Fitz the fifth and Mr. Nicholas Slanning about whether a parcel of land Fitz held of Slanning was freehold, or copyhold. They began their quarrel at the table and Fitz attacked Slanning with his dagger. Separated by their friends, and with the quarrel supposedly patched up, Slanning left. On his way to Bickleigh, he dismounted and sent his man to lead the horses around a rough descent while he walked alone through a field.
>
> Meanwhile, Fitz, with four men attending him, had followed. Though Slanning was on the defensive, some progress was made in talking the matter over and they were putting up their swords when Fitz' man, named Cross, chided him for giving up the quarrel.
>
> Impulsively, Fitz again drew his sword. With this sudden development, Slanning stepped backwards. His spur caught in the ground and he was "foully killed" by Fitz.
>
> John Fitz the fifth, aged twenty-four, having committed his first murder, fled to France. He stayed there until December 1599 by which time his mother and his wife had obtained a pardon for him from Queen Elizabeth.
>
> However, Slanning's widow sued for damages and won the verdict. (1) pg 17

But what about the knighthood? John Fitz the fifth was a pardoned murderer. In her book Getchell relates:

> Upon James I's coronation in 1603, John Fitz [the fifth] was among those knighted, not because of any special service to the Crown but probably because of his family connections and properties. (We have also read that King James

I raised needed revenues at that time by selling knight-
hoods.) (1) pg 17

Despite having survived the results of his actions, John the fifth did not
change his ways, He continued his riotous living and at least one account
says that it got bad enough that his wife took their daughter and returned to
her father's home.

So it appears that Sir John Fitz the fifth was summoned to appear in
court in London, perhaps to answer charges brought by the Slanning children or perhaps charges brought by his wife's father in an attempt to protect
his daughter's interest in the Fitz properties, or perhaps both.

At any rate, Getchell describes it this way:

> Sir John [the fifth] set out on horseback with one ser-
> vant and was evidently most fearful of what would befall
> him in London, or on the way there. 'Dissipation had
> weakened his mind and shattered his nerves' [according to
> Baring-Gould]. He was apparently afraid he was going to
> be waylaid and murdered. At Kingdom-on-Thames he dis-
> missed his servant, having become suspicious that he was in
> league with 'his enemies', and rode off alone into the night.
>
> In Twickenham, at a small tavern called the 'Anchor' he
> demanded admittance at two A.M. The owner, one Daniel
> Alley, most reluctantly admitted him and gave him his own
> bed as none other was available. Mistress Alley went to sleep
> with the children. For the rest of the night they listened to
> Fitz tossing and turning and crying out about his enemies.
>
> At dawn Daniel Alley was about to go out to mow a
> meadow with his neighbor. His wife begged not to be left
> alone in the house with Fitz not yet gone. Alley and his
> friend went up to the chamber door and conversed together
> about what to do. Sir John [the fifth], hearing them, rushed
> out in his nightgown, sword drawn. Thinking his 'enemies'
> had surrounded him, he fell upon his host — and killed
> him. He then wounded Alley's wife with his sword. With
> daylight coming, he discovered what he had done, and in

despair stabbed himself in two places. A surgeon bound up his wounds, but he tore off the bandages and bled to death.

Daniel Alley was buried 8 August 1605. Sir John Fitz [the fifth] was buried 10 August 1605. (1) pgs 17–18

Were the stars right?

Mary the sixth, being the only child of John Fitz the fifth, was the only heiress to her father's considerable estate. In those days in England, women didn't have much to say in their lives including their marriages. Mary was married to Sir Allen Percy when she was only 12 years old. Percy was 31 but the marriage was short lived as he died 3 years later leaving Mary as a widow at age 15. She then married Hon. Thomas Darcy who was her same age but he died a few months after their marriage. Mary, the sixth, married a third time to Sir Charles Howard. They had two daughters, Elizabeth the seventh and Mary the seventh. Howard died after ten years of marriage and Mary married a fourth time to Sir Richard Grenville. One has to wonder about Grenville's interest in the marriage because it rapidly fell apart over money issues. At one point Grenville was sent to Debtor's prison from which he escaped and promptly fled. While he was gone Mary got a divorce. The original Fitz properties had gone back and forth between Grenville and Mary the sixth but she ended up with them.

When Mary the sixth died she left small legacies to her daughters Mary the seventh and Elizabeth the seventh. Fitzford and all her estates she left to her first cousin, Sir William Courtenay. The Fitzford estate was sold to the Duke of Bedford by the Courtenays in 1750 and the Fitz family no longer had an interest in Fitzford.

Apparently, Mary the sixth suffered from a less than stellar reputation, perhaps inherited from her father.

> A local story tells that Mary the sixth as a punishment for mysteriously eliminating her first three husbands has been doomed to traveling from the gateway of Fitzford to Okehampton Park every night in a coach made of human bones and followed by a bloodhound. The hound is to carry back one blade of grass each night and this must continue until every blade of grass in the Park has been picked. An excellent ghost story, but certainly fictitious. Another story

says Mary the sixth herself appears in the shape of a black hound. (1) pg 20

As one of the authors I (Clarence), was rather intrigued that since I enjoy a glass of beer at 5:00 each evening, the Fitz family name in America might had descended from the Sir John Fitz the fifth of Fitzford who obviously enjoyed a glass of beer, or probably too many glasses. But that cannot be the case because Sir John the fifth and his wife only had the one daughter Mary the sixth who could not have passed on the Fitz family name.

While the evidence is far from solid, the Fitz family name in America might have descended from Walter Fitz the second who was the second son of John Fitz the first, Bencher of Lincolns Inn. Walter was born about 1455 and married Mary Sampson. Robert Fitz the third was their second son.

CHAPTER 2

THAT LAND CALLED AMERICA

"In 1452, Catholic Pope Nicholas V released an order that authorized the Christian nations to establish ownership to any lands that they encountered or discovered that were not already inhabited and claimed by other Christian rulers, and to take possession of the land, and property, and the non-Christian inhabitants (called pagans or heathens) of that land. This order is referred to as the 'papal bull' which simply means an order from the pope. Thus, there was authorization from the highest religious leader of the day to take possession of any lands and peoples that they encountered who were not Christian, claiming ownership of said lands and people, usually for the leader of the nation that they served, and all done under the umbrella of Christendom. "*...And the Mille Lacs who have no Reservation...*" Clarence Ralph Fitz - 2016

Using the papal bull as authorization, English settlers made two unsuccessful attempts at establishing a colony in what is now North Carolina. Both failed, the second group just completely disappearing. While this was happening, Chief Powhatan inherited leadership and built a chiefdom of some thirty tribes in what is now eastern Virginia.

In 1603 after the death of Queen Elizabeth I, James VI of Scotland assumed the English throne as King James I. That same year an English voyage was made to the Chesapeake Bay. Captain Gilbert and four others went ashore and were killed by the Indians.

On April 10, 1606, King James I granted land in North America to two competing branches of the Virginia Company, the Virginia Company of London and the Virginia Company of Plymouth.

The Search for the Real Robert

Virginia

On December 19 or 20, 1606, three ships, the Susan Constant, the Godspeed and the Discovery, left England with 104 men and boys aboard, bound for America with the intent of establishing a colony that would produce profits for the Virginia Company, and with the expectation of finding gold and silver in abundance. The Virginia Company of London had been chartered by King James I for that very purpose. Author David Price in his book *Love & Hate in Jamestown* writes "most of them with pure hearts and empty heads, expecting to find riches, welcoming natives, and an easy life on the other shore." (3) pg 13

One of the Jamestown ships — historical recreation.
Photo by author, 1963.

On board were men who had been chosen because of their family connections to the ruling class, or the land holdings of their families, or that they had invested in the venture with the Virginia Company; "gentlemen" they were called. Edward Maria Wingfield with his two servants, was there because he had invested in the Virginia Company venture as well as his father having been a godson of Queen Mary. He had studied law at Lincoln's Inn some many years after John Fitz the first was bencher of the Inn. Wingfield was described as "a covetous haughty person" (3) pg 16 which would translate into laziness.

Also on board were laborers, bricklayers, planters, drummers and other "commoners" as they were called including John Smith. John Smith can be described as a cunning, street-wise and outspoken man, who had had sea adventures previous to this one and who was not intimidated by the so-called "gentlemen" on board.

Jamestown ship — historical recreation.
Photo by author, 1963

The sailing ships of that day were completely reliant on the wind for propulsion so if the wind was blowing the wrong direction, they had no choice but to drop anchor and wait for the wind to blow in the direction they wanted to travel. That is exactly what happened a few days after the three ships left the London harbor. So, for a full month the ships sat anchored with England still in view until the wind shifted. Some of the 71 passengers on board the lead ship Susan Constant, became frustrated at not making any progress in bad weather. Wingfield and some others of high rank wanted to turn back to the nearby comforts of home. John Smith, being a seasoned seaman and not one to defer to the upper class when he knew better, argued against them. By early February they were on their way so tempers calmed but tempers flared again when the ships stopped in the Canary Islands for fresh water and food and at that point rank prevailed and John Smith was under arrest. As a result, the three ships stopped in the West Indies where Captain Newport intended to execute John Smith. As luck would have it, others would intervene and John Smith was spared. After another month the three ships arrived in the Chesapeake Bay on April 26th, where they opened the sealed orders from the Virginia Company. To their chagrin, Captain John Smith was one of those named to be on the governing council, so they apparently felt obligated to release him from prison but not quite ready to allow him a seat on the council. Edward Maria Wingfield was chosen as president of the council. On April 29, 1607, as per Pope Nicholas V's instructions, they erected a cross at Cape Henry claiming the land for King James I.

Their first encounter with the "welcoming natives" that they had been told about back in England came when a scouting party was attacked as they returned to their ship resulting in one colonist being shot by an arrow through both of his hands and another with two arrow wounds to the body. By May of 1607 they had chosen a triangular area bordering on the James river about fifty miles up the river. And while the area is not technically an island it is referred to as James Island, ideal by being surrounded on three sides by water and therefore easy to defend against the Spanish, quite far inland with water deep enough for their ships, not inhabited by Indians and with a narrow land bridge to the mainland.

Captain Newport, Captain John Smith and others spent six days exploring the river up to the falls located near what is now Richmond. Along the way they encountered numerous Indians. After all they were about to establish a colony on the hunting grounds of the Powhatan Indians.

This was to become the first permanent settlement in North America, but it almost didn't happen. Within a few days after arrival, they were attacked by about two hundred Powhatan warriors during which one settler was killed and several were wounded. Because they had been told by the promoters that the Indians would welcome them and that they should treat the Indians with kindness, they had only built a makeshift wall between trees as protection against attack and their guns were still in crates. They immediately started building a more substantial fort. A few days after the fort was completed two of their ships set off for England for more supplies. With them they took mineral samples, hoping to find gold.

By June 10th Captain John Smith was finally sworn in as a member of the council, but all was not going well and the President of the council, Wingfield, who had been caught saving the good food and drink for himself and giving everyone else inferior and scarce food, was removed, imprisoned, and replaced by John Ratcliff. The orders from the Virginia Company were specific in ordering that no colonist could return to England without permission of the president and the council and that no letters containing bad news could be sent back to England, all elements of martial law, whether they were legally empowered with issuing martial law or not. But there seemed to be no instructions on how the colony was to be fed except from the supplies brought on the ships. Just search for gold. A recipe for disaster. John Smith wrote, "At this time our diet was for most part water and bran, and three ounces of little better stuffe in bread for five men a meale, and thus we lived neere three months: our lodgings under boughs of trees, the salvages [Indians] being our enemies, whom we neither knew nor understood; occasion I thinke sufficient to make men sicke and die." (3) pg 52 Smith continued writing, "in such despaire as they would rather starve and rot with idleness, then be persuaded to do anything for their owne reliefe without constraint." (3) pg 55

The colony was about to enter a very difficult period. During the summer and early fall of 1607 half of the settlers would die of disease and starvation, partly because James Island that they had chosen for their colony is very swampy, a perfect place for disease transmission and because they had not seen fit to dig a well for fresh water so they were drinking the brackish and salty river water. By the end of 1607 there were 38 colonists left. But why they did not catch fish, shoot deer and birds and take advantage of the bounty of the land is incomprehensible. Was it just that the leaders they chose were all gentry who had no experience with working?

Instead in desperation they sent John Smith to trade with the Indians for food. Captain John Smith and two others traveled up the Chickahominy River in search of food. Instead, they encountered Powhatan's brother who captured the three settlers. Having killed the other two, John Smith was taken to Chief Powhatan on December 29, 1607. While some historians doubt what we are about to describe, we have come to believe that the story of Powhatan's daughter Pocahontas saving John Smith's life really happened. The copious writings of John Smith and other colonists of the time, all well researched by David Price in writing *Love and Hate in Jamestown* is convincing. Most people who try to tell stories that are not true have a hard time keeping their story straight when told multiple times. Such was not the case with John Smith. We will tell the story, as some of us were taught in grade school, and as related by Chief Justice of the Supreme Court John Marshall in 1804:

> There he was doomed to be put to death, by laying his head upon a stone, and beating out his brains with clubs. He was led to the place of execution, and his head bowed down for the purpose of death, when Pocahontas, the king's darling daughter, then about thirteen years of age, whose entreaties for his life had been ineffectual, rushed between him and his executioner, and folding his head in her arms, and laying hers upon it, arrested the fatal blow. Her father was then prevailed on to spare his life. (3) pg 68

When the second supply ship returned from England it carried very little food, having expected the kindly Indians to supply them with food. John Ratcliff who had replaced Wingfield as council president wrote to the Virginia Company saying, "Though there be fish in the sea, fowls in the air, and beasts in the woods, their bounds [territories] are so large, they so wilde, and we so weake and ignorant, we cannot much trouble them." With that leadership the colony faced another hard winter by December of 1608. Once again John Smith went into action. With a combination of trade for items the Indians relished and of threats and persuasion Smith managed to get food from several of the tribes but when he visited Powhatan, unbeknown to him, he was about to walk into a trap that would have meant his life. Once again at great risk, Pocahontas was able to warn John Smith that he was entering a trap, and with that knowledge Smith was able to escape the trap. By threats and persuasion Smith got enough food to get the colony through until harvest.

By September of 1608, enough of the council members who had sided with and protected the gentry had died that John Smith was elected president. Upon assuming leadership, he spoke to the colonist saying,

"I speake not this to all of you, for divers [several] of you I know deserve honour and reward, better then is yet here to be had. But the greatest part must be more industrious, or starve, how ever you have been heretofore tolerated by the authorities of the councell ... You see now that powers resteth wholly in my selfe; you must obey this now for a law, that he that will not worke shall not eate (except by sickness he be disabled) for the labours of thirtie or fortie honest and industrious men shall not be consumed to maintaine an hundred and fiftie loyterers ... there are now no more counsellors to protect you ... (3) pg 108

Smith's tactic worked and within three months twenty houses were built, a well was dug for clean fresh water, thirty or forty acres were planted and

Jamestown houses — historical recreation. Photo by author, 1963.

a checkpoint was set up at the land bridge to the mainland to prevent raids and stealing. In the spring it was discovered that much of the stored grain had been destroyed by the rats that had made the journey to Jamestown on the ships along with the colonists. In answer, Smith as council president, in addition to sending parties out in search of food was trying to become established in satellite settlements but with mixed results. September of 1609 saw a failed venture when 60 men made their way down river in an attempt to purchase an island. They were in Nansemond Indian territory and the mission went wrong with the chief's son wounded, two English settlers killed and the Indian village burned and ransacked. Captain West was sent upriver to the falls to establish a settlement but met Indian resistance. John Smith himself, tried to purchase Powhatan's town but that went nowhere.

In spite of some failures, under Smith's leadership the colony was finally starting to function but on May 23, 1609 the Virginia Company received a new charter whose terms were that the colony would no longer be ruled by a council but by a single governor. Smith would have been a natural for the position of governor but politics got in the way and Sir Thomas Gates would be the governor. Within a few days nine ships bearing 500 to 600 colonists and supplies left England for the colony only to be hit by a hurricane. One ship, the flagship *Sea Venture* was seriously damaged and limped into Bermuda with all 150 of its passengers including Governor Thomas Gates and John Rolfe. The remaining supply fleet arrived at the colony in August with about 300 new settlers but with few supplies surviving. Unfortunately for John Smith, many of his adversaries were again arriving to oppose how he was leading the colony. Smith insisted that he was still the president until a new charter could be produced or his term ended and the charter was in Bermuda on the remains of the *Sea Venture*. Unfortunately, on one of his excursions to secure satellite settlements his powder bag which he car-

Canon in position — historical recreation. Photo by author, 1963.

ried on his belt ignited and seriously burned him. In September, with Smith at the end of his term and Governor Gates still stranded in Bermuda, George Percy was elected council president. At about this same time Chief Powhatan moved his village further away from the colonists. In October one of the ships from the third supply trip returned to England to report the loss of the *Sea Venture* and most of the supplies. Captain John Smith joined the return voyage to England to recuperate from the suspicious gunpowder explosion. It is interesting to note that Pocahontas was told that John Smith was dead which apparently resulted in her turning her back on the colonists.

Without Smith there was no effective leadership and in short order the colonist returned from their outposts and lived at the fort in fear and without food. The Powhatan Indians took this opportunity in November to attack with devastating results. Three hundred colonist were trapped inside James Fort with few provisions but it would have been much worse had they not been warned of an imminent attack. That winter was the "Starving Time" and colonists resorted to eating their horses, their dogs and cats, any rats, mice or snakes they could capture and even in a few cases, their deceased comrades. When John Smith left to recuperate there were about 500 colonists at the colony. Six months later there were only sixty. It didn't help that Francis West and thirty-six colonists had stolen one of the ships and sailed away with a ship full of food.

By early May there were only 60 survivors at the fort and perhaps because of this reduced threat Powhatan stopped his raids. About that same time the survivors from the *Sea Venture* arrived back at the colony on two ships, the *Deliverance* and the *Patience*, that they had built while stranded

Jamestown Fort — historical recreation. Photo by author, 1963.

on Bermuda. Sir Thomas Gates was on board and now ready to assume his role as governor. Governor Gates tried to bring some order to the colony by implementing martial law. He issued "The Laws Divine, Moral and Martial" with strict codes of behavior and severe punishment for offenders. But within two weeks he decided to abandon Jamestown due to lack of supplies. As luck would have it the departing colonists met Lord De La Warr with supplies and as governor he ordered them to return to the fort. Arriving back at the fort, Gates put the colonists to work cleaning up the fort, repairing the church and building new buildings.

The sparring with Chief Powhatan and his tribes continued through 1610 with loss of life on both sides. In July Governor Gates sent Chief Powhatan an ultimatum: return all English captives and stolen weapons or the settlers will take them back. Chief Powhatan replied with his own ultimatum: either stay in your fort at Fort James or leave Virginia. On March 28, 1611 both Gates and De La Warr returned to England leaving George Percy in charge. Due to continuing disease, there were only 150 settlers remaining.

In May of 1611 Sir Thomas Dale who was to be the new deputy governor arrived from England with 300 men and in August Sir Thomas Gates returned with 280 new settlers. Later that year Dale led 250 men up the river to establish a new settlement near the falls which would be named Henrico probably in an effort to get away from the swampy disease producing and dangerous Jamestown. But they did not get away from the Indians.

By 1612 King James I not only gave the Virginia colony more self-governance but extended its boundaries to include Bermuda. That same year John Rolfe exported his first crop of improved tobacco. He had brought seeds from Bermuda that produced a much sweeter tobacco than the Virginia variety used by the Indians.

In April of 1613 Chief Powhatan's daughter Pocahontas was kidnapped by Captain Samuel Argall and taken to Jamestown for ransom. Her father released several captives and offered corn but the colonist insisted on return of their weapons and tools that had been stolen from the fort. At some point Pocahontas was moved to the settlement at Henrico. By nearly a year later Chief Powhatan hadn't responded to the colonist's ransom demand so a war party of 150 colonists proceeded into Powhatan territory and fighting pursued. But while the ransom demands were being pursued, John Rolfe and Pocahontas were falling in love and they were married on April 5, 1614 following Pocahontas having been baptized and christened "Rebecca." The marriage of Rolfe and Pocahontas effectively ended what is called the First Anglo-Powhatan War, and a son, Thomas Rolfe was born on January 30, 1615. By 1616 there were six English settlements in Virginia with 351 settlers. In May of 1616 the Rolfe family, John, Pocahontas and Thomas left for England accompanied by several Virginia Indians. John Smith was in England at this time and when he heard that the Rolfe's were coming, he wrote to Queen Anne asking that Pocahontas be received as a royal visitor. Among other things he wrote that Pocahontas had saved his life twice: "At the minute of my execution, she hazarded the beating out of her own braines to save mine; and not only that, but so prevailed with her father, that I was safely conducted to James Towne ...." He continued telling the Queen that she had also saved him from ambush in January 1609. (3) pg 173 In March of 1617, just as they were about to leave for Virginia, Pocahontas died. John Rolfe returned to Virginia but he left his son to be raised in England.

By November 18, 1618, both Chief Powhatan and Governor De La Warr having died, the Virginia Company issued what was called the Great Charter which lifted martial law and authorized a General Assembly in which white men of property were authorized to choose representatives and make laws for themselves. The Great Charter established the "headright" system which gave fifty acres of land to anyone who paid their own passage from England to Virginia and an additional fifty acres if they brought someone with them. The second person usually worked as an indentured servant to pay off their debt. Those who had arrived there before 1616, called "Ancient Planters" were each given 100 acres.

The first General Assembly met from July 30, 1619 to August 14, 1619 with the first law they passed requiring that tobacco not be sold for less than three shillings per pound. That same year the first Africans arrived in Virginia, having been stolen off a Portuguese slave ship and sold off an English ship. Slavery was not an established thing in Virginia at that time so at least some of them were treated like indentured servants and gained their freedom just like white indentured servants.

Finally, by November of 1619 the Virginia Company came to the realization that if their colonies were to grow, they needed women, because as of the 1620 census which had counted 1194 colonist, 898 were men with only 141 women, 192 children and 4 Indians. So, by May 90 women would arrive to marry the planters. But there was a catch: their new husband was required to pay their passage. And by fall of 1621 four ships arrived bringing a total of 57 more women. The cost to their new husbands was set at "120 weight [pounds] of the best leafe tobacco for each of them." But when the price of tobacco dropped the price was raised to 150 pounds.

By 1622, relations with the Indians had been good for a while but the Indians were plotting. On March 22, 1622, the Indians under Opechancanough attached the colonist with a vicious attack. By mornings end 347–400 English were dead out of a population of 1240 before the attack. Jamestown itself had been warned so was prepared and was largely spared. When the news reached England in mid-July the English ideology of coexistence and assimilation ended. The Virginia Company was now at war and was prepared to fight. The company urged the colonists to undertake "perpetual warre without peace or truce." (3) Price pg 215 This massacre was the beginning of the second Anglo-Powhatan War which would last about ten years.

With war with the Indians and the need for sentries there was very little harvesting done that fall and the winter of 1622 to 1623 was another "Starving Time" when death from starvation, disease and Indian raids would reach another 500 or so colonists. A letter sent home in April of 1623 described the situation, "To write of all crosses and miseries which have befallen us at this time we are not able: The Lord hath crossed us by striking most of us with sickness and death." (3) Price pg 216

The first Fitz that we find in the records is recorded in the "List of Names of the Living in Virginia, February the 16th, 1623." Under the heading "In James Island" is listed Robert Fitts and under the heading "Over the River" is listed Thomas Fitts and Alice Fitts. (from Hotten pg 178) by Getchell (1) pg 40.

On the "Muster Rolls of Settlers in Virginia" from January 23, 1624 on "James' Island" are listed Robert Fitt, came in the *George* and Ann, his wife came in the *Abigaile*. Listed under property is: Corne, 1 barrell; Pease, 1½ bushels; Peeces, 3; Sword, 1; Swine, 1; and 5 piggs; House, 1." Indicating that Robert was probably a planter (farmer). (1) Getchell pg 40.

One must ponder what life was like in England for the commoners and what would entice men and later women to risk all in America. Was life that wretched in England or was the selling job by the Virginia Company that superb. Price says in his book, *Love and Hate in Jamestown* that "The company continued to recruit new settlers with the promises of ready wealth to be had thanks to the labors of those now dead. What time could be more opportune, many of the newcomers must have reckoned." (3) Price pg 217 A rather grim advertisement but something caused settlers to continue coming to Virginia.

Of note is an entry in the *Virginia Historical Magazine* which quotes the Minutes of the Council and General Court,

> 9th of May 1625: Richard Kingsmill sworn and examined sayeth [that] on Thursdye last paste Robert Fytts was disordered in drinke not being able to goe home contrary to the Proclamation made against drunkennes... And further sayeth [that] John Radishe Caryed over Sr George Yardley his servant to his house at vnsesonable tyme of the night and there gave them Entertainmt' & made them drunke and the next dye drinke likewise to Robert Fytts, wherwth he made himself drunke: [that] is ordered [that] Robert Fitts beinge disordered in drinke shall pay fortie shillings, according to the p'clatione. (1) Getchell pg 40

Ready wealth? Here is a situation that begs the question:

> To the right honorable the Lords and others of His Majestie's most honorable Privie Council. The humble petition of Robert ffitt, Anne his wife and Alice Harris a poore widdowe sheweth that whereas ye poore petitioners having been planters in Virginia for the space of 14 years paste and latelie brought over 16 hogsheades of tobaccoe nowe in the custodie of the ffarmers of the Customes in London for want of payment of Custons and freight which is soe great that the said tobaccoe proving not according to exportation is not able to beare the said Comoditie, being not worth the charge. And the said Robert ffitte beinge now in prison by reason of the stope of the said tobaccoe and ye other petitioners havinge five children on their hands and nott meanes wherewith to give them bredd. Ye petitioners doo nowe humblie beseech your hoborables to graunte them their warrante for the free

discharge of the Custome and Imposte of the said Tobaccoe towards theire presente release, and furtherance of theire retourninge to the plantation againe. And your petitioners with theire children will daylie pray for your honorables. Endorsed 1628 [maybe 1638] (1) Getchell pg 41

Two years after the massacre, the Virginia Company was dissolved by King James and Virginia became a royal colony. David Price, author of *Love and Hate in Jamestown* led us to believe that had it not been for John Smith, Pocahontas and John Rolfe, Jamestown would likely not have survived. What did he mean? It was John Smith's skill in negotiating with the Indians for food that kept the colony from all starving. And it was Pocahontas's admiration for John Smith that kept him from being killed and then later her marriage to John Rolfe that brought some period of peace. John Rolfe by bringing tobacco seed from Bermuda that resulted in a superior tobacco was the beginning of Virginia's tobacco industry. But by the time the last of that trio John Smith died in 1631, the colony, while it still had challenges to face, was well on its way to survival.

Our purpose with this section is not to produce a genealogy, but to prove that the Fitz family was among those that established the first permanent colony in America. That we think we have accomplished.

Massachusetts

In 1617, John Smith would finally get his chance to lead an expedition with the Virginia Company of Plymouth to New England, or so he thought. As the three small ships waited for friendly winds, a wait that extended for three months, the window for such a voyage was over and his last chance to lead a voyage disappeared.

While the colonists who ventured to Virginia were most likely lured primarily in search of gold and silver and other forms of wealth that the promoters had essentially promised them, the colonist who headed for New England were in search of religious freedom. They were the English Separatists, a radical Puritan faction, that King James I disliked adamantly and for whom he was making life intolerable.

In 1620 the Separatists obtained a patent from the Virginia Company of London to establish a plantation on the Hudson River near what is now New

York City. John Smith campaigned to be named the leader of this voyage but instead they chose Miles Standish. But while they chose not to give Smith the leadership role, they knew that he had previously explored the area, so they didn't take Smith but they took his maps and the books he had written describing the area to which they were headed. And so, the *Mayflower* with 100 English Separatist men and women called the Pilgrims set sail for America. While the *Mayflower* was headed for the Hudson River, due to storms and rough seas they ended up in Cape Cod in late December and chose Plymouth Rock as the spot for the first permanent settlement in New England (3) Price pg 224.

We are all at least somewhat familiar with the Pilgrim's first Thanksgiving in the fall of 1621 and the story of the joint feast with the colonists and the Indians. It probably was a celebration of their harvest but most likely it was also a celebration of their newfound freedom of religion without the harassment of King James I. While the Plymouth colony would never develop a robust economy, due to agriculture, fishing and trade, the colony was self-sufficient within five years of its founding, only to be overshadowed by and eventually absorbed by the Massachusetts Bay Company in 1691.

The Massachusetts Bay colony was established in 1630 and encompassed the area to the north of Plymouth around what is now Salem and Boston. And some distance north of Salem was Salisbury where Robert Fitts and his wife Grace D. were original settlers in 1639. (1) JHF pg 1 Sylvia Fitts Getchell writes in her book, "We do not know the parents, nor the birth dates, nor the birth places of the three Fitts' who emigrated from old England to New England. There were two brothers and [probably] one sister. Robert of Ipswich and Salisbury, MA, … Bridget of Newbury and Gloucester … and Richard of Ipswich and Newbury,MA," all areas in relatively close proximity to each other.

Getchell also writes, "There is certainly no question in the writer's mind but that Robert Fitts, his brother Richard, and his possible sister Bridget were all born in England." "As Grace (Lord) Fitts probably came from the Sudbury, County Suffolk area, and as she and Robert were in all probability married in England, it would seem a reasonable supposition that Robert Fitts was also from that area. There is a persistent family tradition that he came from Devonshire.

While not discounting any of the possibilities, it seems that that belief could have arisen simply from the knowledge of the existence of the old family seat of Fitzford in Tavistock. As the somewhat cosmopolitan descendants of bencher, John Fitz" [the first] "of Devon and London, owned land in other parts of England, is it not possible that some (?) of the branches in Suffolk, Norfolk, etc., originally had ties with the old Devon family? The reader may note that the Letheringsett, County Norfolk, family bore virtually the same coat of arms as did the Devon family." (1) Getchell pg 45

James Hill Fitts says in his book, *Genealogy of the Fitts or Fitz Family in America,* "The American ancestor of the family of Fitts or Fitz, was Robert, who, with his wife, Grace D., was among the original settlers of Salisbury, Mass." (4) pg 1 Robert and Grace had a son Abraham. Robert Fitz was a planter, a farmer, and apparently a quite successful one as evidenced by his last will and testament.

Robert Fitz in America

The second son of John Fitz the first, bencher at Lincoln's Inn was Walter. The second son of Walter was Robert. Was Robert, the second son of Walter, the Robert that immigrated to Jamestown, Virginia in about 1623? Or was Robert, the second son of Walter, the Robert that settled in Salisbury, Massachusetts in 1639? Or were they the same Robert? We can conclude, we believe, that Robert of Virginia was the first Fitz to settle in America and we can conclude, we think, that both Roberts were planters [farmers]. It also seems evident that things were not going well for Robert in Virginia as evidenced by not having enough money to pay the duty on the tobacco he had grown and perhaps by his problem with liquor. Might he have decided to abandon Virginia and give it a try in New England? But there are some issues with that theory. First is that Virginia Robert's wife was Ann and Massachusetts Robert's wife was Grace. That is not impossible to explain with the death rate in Virginia, but what about the children. We know that in the tobacco duty court case reference was made to five children under the care of Robert's wife Ann and the widow Alice Harris, but we don't know who the children's parents were. We do know that Massachusetts Robert had a son, Abraham. We also know that Robert of Virginia was struggling to become successful as a farmer and Robert of Massachusetts appeared to be very successful. While none of these issues is definitive and none of them would be impossible if we had the facts, we tend to agree with Sylvia Fitts Getchell that Robert of Virginia and Robert of Massachusetts were different people.

Cousin Carrie (Fitz) Lambert seems to think they were the same person. In a 1977 genealogy report entitled *Fitz Family History and Genealogy* she writes, "Among the earliest records of the name in America, was that of his [John the first] grandson, Robert, who emigrated to the Virginia colony in 1622, but later moved to Salisbury and then to Ipswich, Mass." (5) Lambert pg 2. She may be right.

Suffice it to say that the Fitz family name first arrived in America in 1622 or 1623 at Jamestown, Virginia, the first permanent colony in America.

CHAPTER 3

GO WEST YOUNG MAN, GO WEST

In Chapter 2 we briefly discussed the "papal bull" which established the "Doctrine of Discovery." The Doctrine of Discovery set the stage for what is called Manifest Destiny.

> Using the Doctrine of Discovery as a foundation, it is easy to understand how settlers in the 1800's believed, rightly or wrongly, that it was their manifest destiny to push westward all the way to the Pacific, taking possession of and settling the land and saving souls by converting the Indians to Christendom, while at the same time using whatever method they deemed necessary to move the Indians out of the way so settlement could continue unimpeded. They were simply doing what God, through the church that spoke for God, had asked them to do. (1) Fitz

Perhaps though, the fact that large families were common in the era we are writing about, and taking into consideration that most farms in the beginning were focused on growing enough produce to feed and clothe their own families with hopefully some to market so that items not produced on the farm could be purchased, as young men and women reached the age at which they were ready to start their own careers and families they really had no choice but to push further west where land was available, and Manifest Destiny gave them license, perhaps unknowingly, to do so.

The further back we try to trace the family, the murkier the record becomes. Henry Fitz, probably born Henry Fitz Randolph (according to Max Fitz through Steva Lynn Davis), and confirmed it would seem by the New Jersey marriage records that show Henry's son John Fitz Randolph marrying Elizabeth Dunn on November 24, 1807. This being the case, both father and son apparently dropped the Randolph from their names. It also seems to indicate that the Henry Fitz we are talking about here probably has no connection to the Robert Fitz discussed in Chapter 2.

Great-great-great grandfather Henry Fitz was born about 1761 in Fayette County, Pennsylvania. He and his wife Jane farmed in Fayette County on a farm he had purchased. Louise Fitz Walters writes in *The (Henry) Fitz Family*,

> Henry Fitz was born about 1761, and about 1785 he moved into Fayette Co., in southwestern PA. The land in this area had been purchased from the Indians in 1768, but continuing raids by Indians who were unhappy about the sale made it a very unsafe place for whites. Settlement was slow until after the end of the Revolutionary War, when settlers flocked in and the Indians moved on into Ohio.
>
> In 1791, Henry Fitz became one of the founders of a Baptist Church at Big Redstone, one of the first churches west of the Alleghenies. Henry's wife was Jane and their children were John, David, Henry, Mary, Elizabeth, Eliza, Jane and Lydia Ellen. Henry died June 16, 1837.

The Fitz ancestor who brought the authors branch of the Fitz family to Iowa was Thomas B. Fitz who was the second son of our great-great grandfather John Fitz born February 2, 1787 in Burlington, Western New Jersey and Elizabeth Dunn whom he married on November 27, 1807. Louise Fitz Walters in *The (Henry) Fitz Family* writes,

> ...They raised a family of twelve children: Henry, Thomas, Melinda, Mary, Wiley, Alvin, Jane, William, David S., John, Samuel and James M. Some of these children were born in PA, and after about ten years of marriage, the family moved to Muskingum County, in southeastern Ohio. This

must have been a very difficult trip, as all of their belongings had to be moved by wagon and the road in some places was only a trail wide enough for the wagon. But Ohio was new country, and very fertile. On John's farm he had hogs, cattle, sheep, chickens, geese, horses and oxen. The family raised its own food as well as raising wheat and flax. Besides the farm implements and tools, there were three spinning wheels for making yarn from flax and wool from the sheep, and weaving loom for weaving the yarn into bed covers and possibly for making clothing.

John was among one of the first deacons of the Adamsville Baptist Church.

Thomas B. Fitz, great grandfather of the authors of this book, was born April 1, 1810, and according to Sylvia Fitts Getchell probably in Franklin, Pennsylvania which is in Fayette County and only 35 or 40 miles from the

Thomas B. Fitz

Ohio border but according to Cousin Carrie (Fitz) Lambert he was born in Muskingum County near Zanesville, Ohio. We don't know what intervened between his birth and his marriage but according to Cousin Carrie Lambert he married Charlotte Bruner, a Pennsylvania Dutch girl in Rafstone Baptist Church in Fayette County, Pennsylvania in 1833. Thomas and his wife Charlotte who was also born in 1810 and according to Getchell born in Ohio, their first two children, Elizabeth Ann who was born in 1834 and John Abraham who was born in 1836 migrated from Ohio to Linn County, Iowa sometime between 1836 and 1840.

With the Louisiana Purchase from France under President Thomas Jefferson of the area that stretched from New Orleans to the Canadian border, the size of the United States essentially doubled. That purchase in 1803 included the entire area that would become the State of Iowa in 1846.

Thomas B. Fitz, research tells us, immigrated with his family to Iowa sometime between 1836 and 1840. That would have been between four years and eight years after the Black Hawk war which pitted a portion of the Sac and Fox Indian tribe under the leadership of Chief Black Hawk against the United States and the fighting force made up of volunteers and soldiers of the regular army. The result of that war which ended in 1832 was that the followers of Black Hawk were either dead or in prison, leaving only the friendly Sac and Fox Indians under Chief Keokuk in what would become the State of Iowa.

Charlotte Bruner Fitz

These were brave and sturdy people, our ancestors, Thomas and Charlotte Fitz. Only a few short years before their arrival in what would become Iowa, there was such fierce fighting between the Sioux Indians and the Sac and Fox Indians that the federal government under President Monroe negotiated a treaty which established a line between the two tribes. Even that didn't stop the fighting so the government in 1830 resorted to buying enough land from both tribes to establish a neutral zone.

When we authors discovered a book, a historical document, not copyrighted which makes it public domain free to be copied, entitled *History of Franklin County, Iowa, a Record of Settlement, Organization, Progress and Achievement*, edited by I.L. Stuart and published in1914, we decided that the history of the area between the purchase of Louisiana from France and the arrival of our ancestors belonged in this book. The following is the section that we deemed appropriate. We have added punctuation to make it more readable, added two pictures and pointed out two misspelled words, but other than that it is copied exactly as it appears in the book.

"Four score years ago all that part of the great and beautiful State of Iowa, of which the county of Franklin is a part, was practically *terra incognita*, a vast wilderness, given over by the Almighty to wild beasts, birds of the air and their masters, the Indians, who roamed the plains and forests at will, claiming and securing an existence from the bounteous hand of nature. Here the deer, buffalo and other fur-bearing

animals found a habitat, and the main streams gave gen-
erously of the palatable fish. The red man had no care for
the morrow. No thought came to him that his possessions
would ever be disturbed by the pale face. So he continued
his dreams.

The hunt was his daily avocation, broken in upon at
intervals by a set-to with a hostile tribe of aborigines, that
was always cruel and bloody in its results and added spoils
to the victor and captives for torture. He knew not of the
future and cared less But the time was coming, was upon
him, when he was called upon to make way for a stronger
and a progressive race of men; when the fair land that was
his birthright and his hunting grounds, resplendent with
the gorgeous flower and emerald sod, must yield to the hus-
bandman. The time had come for the buffalo, deer and elk
to seek pastures new, that the alluvial soil might be turned
to the sun and fed with grain, to yield in their seasons the
richest of harvests.

It is hard for the present generation to realize the rapid pace of civilization on the western continent in the past one hundred years; and when
one confines his attention to the advancement of the State of Iowa in the
past seventy-five years, his amazement is all the more intense. Evidence of
progress are on every hand as one wends one's way across the beautiful state.
Manufacturing plants are springing up hither and yon; magnificent edifices
for religious worship point their spires heavenward; schoolhouses, colleges,
and other places of learning and instruction make the state stand out prom-
inently among her sisters of the great republic. Villages are growing into
towns, and towns are taking on the dignity of a city government, until today
Iowa is noted throughout the Union for the number, beauty and thrift of her
towns and cities. The commonwealth is cobwebbed with her telegraph, telephone and railroad lines, and all these things above mentioned have been
made possible by the thrift, determination and high character of the people
who claim citizenship within her borders.

It is conceded by historians who have given the subject deep thought
and careful research that this country was inhabited by a race of human

beings distinct from the red man. But that is beyond the province of this work. The men and women who opened up the State of Iowa and the county of Franklin to civilization had only the red man to dispute their coming and obstruct their progress; and in that regard something should be recorded in these pages.

So far as the writer can ascertain, the Indians were the first inhabitants of Iowa. For more than one hundred years after Marquette and Joliet had trod the virgin soil of Iowa and admired its fertile plains, not a single settlement had been made or attempted; not even a trading post established. The whole country remained in the undisputed possession of the native tribes. These tribes fought among themselves and against each other for supremacy and the choicest hunting grounds became the reward for the strongest and most valiant of them.

When Marquette visited this country in 1673, the Illini were a powerful people and occupied a large portion of the state, but when the country was again visited by the whites, not a remnant of that once powerful tribe remained on the west side of the Mississippi, and Iowa was principally in the possession of the Sacs and Foxes, a warlike tribe which, originally two distinct nations residing in New York and on the waters of the St Lawrence, had gradually fought their way westward and united, probably after the Foxes had been driven out of the Fox river country in 1846 and crossed the Mississippi. The death of Pontiac, a famous Sac chieftain, was made the pretext for war against the Illini, and a fierce and bloody struggle ensued, which continued until the Illini were nearly destroyed, and their possessions went into the hands of their victorious foes. The Iowas also occupied a portion of the state for a time, in common with the Sacs, but they, too, were nearly destroyed by the Sacs and Foxes and in the "Beautiful Land," these natives met their equally warlike and bloodthirst enemies, the Northern Sioux, with whom they maintained a constant warfare for the possession of the country for a great many years.

In 1803. when, under the administration of Thomas Jefferson, the President of the United States, Louisiana was purchased from Napoleon Bonaparte, Emperor of France, the Sacs, Foxes and Iowas possessed the entire State of Iowa and the two former tribes also occupied most of Illinois. The Sacs had four principal villages, where most of them resided. Their largest and most important town, from which emanated most of the obstacles encountered by the Government in the extinguishment of Indian titles to land in this region, was on the Rock river, near Rock Island; another was on the east bank of the Mississippi, near the mouth of Henderson river; the third was at the head of the Des Moines rapids, near the present site of Montrose, and the fourth was near the mouth of the Upper Iowa. The Foxes had three principal villages. One was on the west side of the Mississippi, six miles above the rapids of Rock river, another was about twelve miles from the river, in the rear of the Dubuque lead mines, and the third was on Turkey river.

The Iowas, at one time identified with the Sacs of Rock river, had withdrawn from them and become a separate tribe. Their principal village was on the Des Moines river, in Van Buren county, on the site where Iowaville now stands. Here the last great battle between the Sacs and Foxes and Iowas was fought, in which Black Hawk, then a young man, commanded one division of the attacking forces. The following account of the battle has been given.

> Contrary to long established custom of Indian attack, this battle was commenced in the daytime, the attending circumstances justifying this departure from the well settled usages of Indian warfare The battlefield was a level river bottom, about four miles in length and two miles wide near the middle, narrowing to a point at either end The main area of this bottom rises perhaps twenty feet above the river, leaving a narrow strip of low bottom along the shore covered with trees that belted the prairie on the river side with a thick forest, and the immediate bank of the river was fringed with a dense growth of willows. Near the lower end of this prairie, near the river bank, was situated the Iowa village. About two miles above it and near the middle of the prairie is a mound, covered at the time with a small clump of trees and underbrush growing on its summit. In the rear of the little elevation, or mound, lay a belt of wet prairie, covered at that time with a dense growth of rank, coarse grass Bordering this wet prairie on the north, the country rises abruptly into elevated, broken river bluffs, covered with a heavy forest for miles in extent and in places thickly clustered with

undergrowth, affording convenient shelter for the stealthy approach of an enemy

Through this forest the Sac and Fox war party made their way in the night and secreted themselves in the tall grass spoken of above, intending to remain in ambush during the day and make such observations as this near proximity to their intended victims might afford, to aid them in their contemplated attack on the town during the following night From this situation their spies could take a full survey of the village and watch every movement of the inhabitants, by which means they were soon convinced that the Iowas had no suspicion of their presence.

At the foot of the mound above mentioned the Iowas had their race course, where they diverted themselves with the excitement of horse racing and schooled their young warriors in cavalry evolutions In these exercises mock battles were fought and the Indian tactics of attack and defense carefully inculcated, by which means a skill in horsemanship was acquired that is rarely excelled Unfortunately for them, this day was selected for their equestrian sports and, wholly unconscious of the proximity of their foes, the warriors repaired to the race ground, leaving most of their arms in the village, and their old men, women and children unprotected

Pash-a-popo, who was chief in command of the Sacs and Foxes, perceived at once this state of things afforded for a complete surprise of his now doomed victims, and ordered Black Hawk to file off with his young warriors through the tall grass and gain the cover of the timber along the river bank, and with the utmost speed reach the village and commence the battle, while he remained with his division in the ambush to make a simultaneous attack on the unarmed men whose attention was engrossed with the excitement of the races The plan was skillfully laid and dexterously executed Black Hawk with his forces reached the village undiscovered, and made a furious onslaught upon the defenseless inhabitants by firing one general volley into their midst and completing the slaughter with the tomahawk and scalping knife, aided by the devouring flames with which they enveloped the village as soon as the fire-brand could be spread from lodge to lodge

On the instant of the report of firearms at the village, the forces under Pash-a-popo leaped from their couchant position in the grass and sprang, tigerlike, upon the unarmed Iowas in the midst of their racing sports The first impulse of the latter naturally led them to make the utmost speed toward their arms in the village, and protect, if possible, their wives and children from the attack of their merciless assailants The distance from the place of attack on the prairie was two miles, and a great number fell in their flight by the bullets and tomahawks of their enemies, who pressed them closely with a running fire the whole way and the survivors only reached their town to witness the horrors of the destruction

Their whole village was in flames and the dearest objects of their lives lay in slaughtered heaps midst the devouring element, and the agonizing groans of the dying, mingled with the hideously exulting shouts of the enemy, filled their hearts with maddening despair Their wives and children who had been spared the general massacre were prisoners, and their weapons in the hands of the victorious savages; all that could be done was to draw off their shattered and defenseless forces, and save as many lives as possible by a retreat across the Des Moines river, which they effected in the best possible manner, and took a position among the Soap Creek hills

The Sioux located their hunting grounds north of the Sacs and Foxes. They were a fierce and warlike nation and often disputed possession in savage and fiendish warfare. The possessions of these tribes were mostly located in Minnesota but extended over a portion of northern and western Iowa to the Missouri river. Their descent from the north upon the hunting grounds of Iowa frequently brought them into collision with the Sacs and Foxes and after many a sanguine conflict, a boundary line was established between them by the Government of the United States, in a treaty held at Prairie du Chien in 1825. Instead of settling the difficulties, this caused them to quarrel all the more, in consequence of alleged trespasses upon each other's side of the line. So bitter and unrelenting became these contests that in 1830 the Government purchased of their respective tribes of the Sacs and Foxes and the Sioux, a strip of land twenty miles wide on both sides of the line, thus throwing them forty miles apart by creating a "neutral ground" and commanded them to cease their hostilities. They were, however, allowed to fish on the ground unmolested, provided they did not interfere with each other on United States territory.

Soon after the acquisition of Louisiana the United States government adopted measures for the exploration of the new territory, having in view the conciliation of the numerous tribes of Indians by whom it was possessed, and also the selection of proper sites for the establishment of military posts and trading stations. The Army of the West, General Wilkinson commanding, had its headquarters at St Louis. From this post Captains Lewis and Clark, with a sufficient force, were detailed to explore the unknown sources of the Missouri, and Lieut Zebulon M. Pike to ascend to the headwaters of the Mississippi. Lieutenant Pike, with one sergeant, two corporals and seventeen privates, left the military camp near St Louis, in a keel boat, with four months rations, August 9, 1805. On the 20th of the same month the expedition arrived within the present limits of the State of Iowa, at the foot of the Des Moines rapids, where Pike met William Ewing, who had just been appointed Indian agent at this point; a French interpreter, four chiefs, fifteen Sac and Fox warriors. At the head of the rapids, where Montrose is now situated, Pike held a council with the Indians, in which he addressed them substantially as follows:

> Your great father, the President of the United States, wishes to be more acquainted with the situation and wants of the different nations of red people in our newly acquired territory of Louisiana and has ordered the General to send a number of his warriors in different directions to take them

by the hand and make such inquiries as might afford the satisfaction required.

At the close of the council he presented the red men with some knives, tobacco and whiskey. On the 23rd of August he arrived at what is supposed from his description to be the site of the present Burlington, which he selected as the location for a military post. He describes the place as "being on a hill, forty miles above the River de Moyne rapids, on the west side of the river, in latitude about forty degrees twenty-one minutes north. The channel of the river runs on that shore. The hill in front is about sixty feet perpendicular, and nearly level at the top. About four hundred yards in the rear is a small prairie, fit for gardening, and immediately under the hill is a limestone spring sufficient for the consumption of a whole regiment.

In addition to this description, which corresponds to Burlington, the spot is laid down on his map at a bend in the river a short distance below the mouth of the Henderson, which pours its waters into the Mississippi from Illinois. The fort was built at Fort Madison but from the distance, latitude, description and map furnished by Pike, it could not have been the place selected by him, while all the circumstances corroborate the opinion that the spot he selected was the place where Burlington is now located, called by the early voyagers on the Mississippi "Flint Hills." In company with one of his men Pike went on shore on a hunting expedition and following a stream which they supposed to be a part of the Mississippi they were led away from their course. Owing to the intense heat and tall grass, his two favorite dogs, which he had taken with him, became exhausted, and he left them on the prairie, supposing they would follow him as soon as they should get rested, and went on to overtake his boat. After reaching the river he waited for some time for his canine friends but they did not come, and as he deemed it inexpedient to detain the boat longer, two of his men volunteered to go in pursuit of them. He then continued on his way up the river, expecting the men would soon overtake him. They lost their way, however, and for six days were without food, except a few morsels gathered from the stream. They might have perished had they not accidentally met a trader from St Louis, who induced two Indians to take them up the river, overtaking the boat at Dubuque. At the latter place Pike was cordially received by Julien Dubuque, a Frenchman, who held a mining claim under a grant from Spain. He had an old field piece and fired a salute in honor of the advent of the first American who had visited that part of the territory. He was not, however, disposed to publish the wealth of his mines and the young, and evidently inquisitive officer obtained but little information in that regard.

Upon leaving this place Pike pursued his way up the river but as he passed beyond the limits of the present State of Iowa, a detailed history of his exploration does not properly belong to this volume. It is sufficient to say that on the site of Fort Snelling, Minnesota, he held a council with the Sioux, September 23rd, and obtained from them a grant of one hundred thousand acres of land.

Before the territory of Iowa could be opened to settlement by the whites it was first necessary that the Indian title should be extinguished and the aborigines removed. The territory had been purchased by the United States but was still occupied by the Indians, who claimed title to the soil by right of possession. In order to accomplish this purpose, large sums of money were expended, warring tribes had to be appeased by treaty stipulations and oppression by the whites discouraged.

When the United States assumed control of the country, by reason of its purchase from France, nearly the whole state was in possession of the Sacs and Foxes, a powerful and warlike nation, who were not disposed to submit without a struggle to what they regarded the encroachment on their rights of the pale faces. Among the most noted chiefs and one whose restlessness and hatred of the whites occasioned more trouble to the Government than any other of his tribe, was Black Hawk, who was born at the Sac village, on Rock river, in 1767. He was simply the chief of his own band of Sac warriors; but by his energy and ambition he became the leading spirit of the united nation of Sacs and Foxes, and one of the prominent figures in the history of the country from 1804 until his death. In early manhood he attained distinction as a fighting chief, having led campaigns against the Osages and other neighboring tribes. About the beginning of the nineteenth century he began to appear prominent in affairs on the Mississippi. His life was a marvel. He is said by some to have been the victim of a narrow prejudice and bitter ill feeling against the Americans.

Chief Black Hawk

November 3, 1804, a treaty was concluded between William Henry Harrison, then governor of the Indian Territory, on behalf of the United States, and five chiefs of the Sac and Fox nation, by which the latter, in consideration of $2,234 in goods then delivered, and a yearly annuity of $1,000 to be paid in goods at just cost, ceded to the United States, all that land on the west side of the Mississippi extending from a point opposite the Jefferson, in Missouri, to the Wisconsin river, embracing an area of 51,000,000 acres. To this treaty Black Hawk always objected and always refused to consider it binding upon his people. He asserted that the chiefs and braves who made it had no authority to relinquish the title of the nation to any of the lands they held or occupied and, moreover, that they had been sent to St Louis on quite a different errand, namely, to get one of their people released, who had been imprisoned at St Louis for killing a white man.

In 1805, Lieutenant Pike came up the river for the purpose of holding friendly council with the Indians and selecting sites for forts within the territory recently acquired from France by the United States. Lieutenant Pike seems to have been the first American whom Black Hawk had met or had a personal interview with and was very much impress in his favor. Pike gave a very interesting account of his visit to the noted chief.

Fort Edwards was erected soon after Pike's expedition, at what is now Warsaw, Illinois, also Fort Madison, on the site of the present town of that name, the latter being the first fort erected in Iowa. These movements occasioned great uneasiness among the Indians. When work was commenced on Fort Edwards, a delegation from the nation, headed by their chiefs, went down to see what the Americans were doing and had an interview with the commander, after which they returned home and were apparently satisfied. In like manner, when Fort Madison was being erected, they sent down another delegation from a council of the nation held at Rock river.

According to Black Hawk's account, the American chief told them he was building a house for a trader, who was coming to sell them goods cheap, and that the soldiers were coming to keep him company — a statement which Black Hawk says they distrusted at the time, believing that the fort was an encroachment upon their rights, and designed to aid in getting their lands away from them. It is claimed, by good authority, that the building of Fort Madison was a violation of the Treaty of 1804. By the eleventh article of that treaty, the United States had the right to build a fort near the mouth of the Wisconsin river, and by article six they bound themselves "that if any citizen of the United States or any other white person should form a settlement upon their lands such intruder should forthwith be removed." Probably the authorities of the United States did not regard the establishment of military posts as coming properly within the meaning of the term "settlement," as used in the treaty.

At all events, they erected Fort Madison within the territory reserved to the Indians, who became very indignant. Very soon after the fort was built, a party led by Black Hawk attempted its destruction. They sent spies to watch the movements of the garrison, who ascertained that the soldiers were in the habit of marching out of the fort every morning and evening for parade, and the plan of the party was to conceal themselves near the fort and attack and surprise them when they were outside. On the morning of the proposed day of the attack five soldiers came out and were fired upon by the Indians, two of them being killed. The Indians were too hasty in their movements, for the parade had not commenced.

However, they kept up the siege several days, attempting the old Fox strategy of setting fire to the fort with blazing arrows, but finding their efforts unavailing, they desisted and returned to their wigwams on Rock river. In 1812, when the war was declared between this country and Great Britain, Black Hawk and his band allied themselves with the British, partly because he was dazzled by their specious promises, but more probably because they were deceived by the Americans. Black Hawk himself declared they were forced into the war by having been deceived. He narrates the circumstances as follows:

> Several of the head men and their chiefs of the Sacs and
> Foxes were called upon to go to Washington to see their

great father. On their return they related what had been said and done. They said the great father wished them, in the event of war taking place with England, not to interfere on either side but to remain neutral. He did not want our help but wished us to hunt and support our families and live in peace. He said that British traders would not be permitted to come on the Mississippi to furnish us with goods but that we should be supplied by an American trader. Our chiefs then told him that the British traders always gave them credit in the fall for guns, powder and goods, to enable us to hunt and clothe our families. He repeated that the traders at Fort Madison would have plenty of goods; that we should go there in the fall and he would supply us on credit, as the British traders had done.

Black Hawk seems to have accepted the proposition and he and his people were very much pleased. Acting in good faith, they fitted out for their winter's hunt and went to Fort Madison in high spirits to receive from the trader their outfit of supplies; but after waiting some time they were told by the trader that he would not trust them. In vain they pleaded the promise of their great father at Washington, the trader was inexorable.

Disappointed and crestfallen, the Indians turned sadly to their own village. Says Black Hawk,

> Few of us slept that night. All was gloom and discontent. In the morning a canoe was seen ascending the river; it soon arrived bearing an express who brought intelligence that a British trader had landed at Rock Island with two boats filled with goods, and requested us to come up immediately, because he had good news for us and a variety of presents. The express presented us with pipes, tobacco and wampum. The news ran through our camp like fire on the prairie. Our lodges were soon taken down and all started for Rock Island. Here ended all our hopes of remaining at peace, having been forced into the war by being deceived.

He joined the British, who flattered him and styled him "General Black Hawk," decked him with medals, excited his jealousy aginst the Americans and armed his band, but he met with defeat and disappointment and soon abandoned the service and returned home.

There was a portion of the Sacs and Foxes whom Black Hawk, with all his skill and cunning, could not lead into hostilities against the United States. With Keokuk, "the Watchful Fox," at their head, they were disposed to abide by the Treaty of 1804 and to cultivate friendly relations with the American people. So, when Black Hawk and his band joined the fortunes of Great Britain, the rest of the nation remained neutral and for protection organized with Keokuk for their chief.

Thus, the nation was divided into the "war party" and "peace party." Keokuk became one of the nation's great chiefs. In person he was tall and of portly bearing. He has been described as an orator, entitled to rank with the most gifted of his race, and through the eloquence of his tongue he prevailed upon a large body of his people to remain friendly to the Americans.

Chief Keokuk

As has been said, the Treaty of 1804, between the United States and the Sac and Fox nations was never acknowledged by Black Hawk and in 1831 he established himself with a chosen band of warriors upon the disputed territory, ordering the whites to leave the country at once. The settlers complaining, Governor Reynolds of Illinois, dispatched General Gaines with a company of regulars and one thousand five hundred volunteers to the scene of action. Taking the Indians by surprise, the troops burned their village and forced them to conclude a treaty, by which they ceded all their lands east of the Mississippi and agreed to remain on the west side of the river.

Necessity forced the proud spirit of Black Hawk into submission, which made him more than ever determined to be avenged upon his enemies. Having rallied around him the warlike braves of the Sac and Fox nations, he recrossed the Mississippi in the spring of 1831. Upon hearing of the invasion, Governor Reynolds hastily gathered a body of one thousand eight hundred volunteers, placing them under Brig. Gen. Samuel Whiteside.

The army marched to the Mississippi and, having reduced to ashes the village known as "Prophet's Town," proceeded several miles up Rock river to Dixon to join the regular forces under General Atkinson. They formed at Dixon two companies of volunteers, who, sighing for glory, were dispatched to reconnortre [sic] the enemy. They advanced under command of General Stillman to a creek, afterward called "Stillman's Run," and while encamping there saw a party of mounted Indians at a distance of a mile.

Several of Stillman's men mounted their horses and charged the Indians, killing three of them, but attacked by the main body under Black Hawk, they were routed and by their precipitate flight spread such a panic through the camp that the whole company ran off to Dixon as fast as their legs could carry them. On their arrival it was found eleven had been killed. For a long time afterward Major Stillman and his men were subjects of ridicule and merriment, which was as undeserved as their expedition was disastrous. Stillman's defeat spread consternation throughout the state and nation. The number of Indians was greatly exaggerated and the name of Black Hawk carried with it association of great military talent, cunning and cruelty. He was very active and restless and was continually causing trouble.

After Black Hawk and his warriors had committed several depredations and added more scalp locks to their belts, that restless chief and his savage partisans were located on Rock river, where he was in camp. On July 19th, General Henry being in command, ordered his troops to march. After having gone fifty miles, they were overtaken by a terrible thunderstorm which lasted all night. Nothing cooled in their ardor and zeal, they marched fifty miles the next day, encamping near the place where the Indians encamped the night before.

Hurrying along as fast as they could, the infantry keeping up an equal pace with the mounted men, the troops on the morning of the 21st crossed the river connecting two of the four lakes, by which the Indians had been endeavoring to escape. They found on their way the ground strewn with kettles and articles of baggage, which in the haste of retreat the Indians were obliged to abandon. The troops imbued with new ardor, advanced so rapidly that at noon they fell in with the rear guards of the enemy.

Those who closely pursued them were saluted by a sudden fire of musketry from a body of Indians who had concealed themselves in the high grass of the prairie. A most desperate charge was made on the four who, unable to resist, retreated obliquely in order to outflank the volunteers on the right but the latter charged the Indians in their ambush and expelled them from the thickets at the point of the bayonet and dispersed them.

Night set in and the battle ended, having cost the Indians sixty-eight of their bravest men, while the loss of the Illinoisans was but one killed and eight wounded. Soon after this battle Generals Atkinson and Henry joined forces and pursued the Indians. General Henry struck the main trail, left his horses behind, formed an advance guard of eight men and marched forward upon the trail. When these eight men came in sight of the river they were suddenly fired upon and five of them killed, the remaining three maintaining their ground until General Henry came up. Then the Indians, charged with the bayonet, fell back upon their main force.

The battle now became general. The Indians fought with desperate valor but were furiously assailed by the volunteers with their bayonets, cutting many of the Indians to pieces and driving the rest of them into the river. Those who escaped from being drowned found refuge on an island. On hearing the frequent discharge of musketry, General Atkinson abandoned the pursuit of the twenty Indians under Black Hawk himself and hurried to the scene of the action, where he arrived too late to take part in the battle.

He immediately forded the river with his troops, the water reaching up to their necks, and landed on the island where the Indians had secreted themselves. The soldiers rushed upon the Indians, killed several of them, took other prisoners and chased the rest into the river, where they were either drowned or shot before reaching the opposite shore. Thus, ended the battle, the Indians losing 300, besides 50 prisoners; the whites but 17 killed and 12 wounded.

Black Hawk with his twenty braves retreated up the Wisconsin river. The Winnebagoes [sic], desirous of securing the friendship of the whites, went in pursuit and captured and delivered them to General Street, the United States Indian agent. Among the prisoners were the son of Black Hawk and the prophet of the tribe. These, with Black Hawk, were taken to Washington, D.C., and soon consigned as prisoners to Fortress Monroe. At the interview Black Hawk had with the president he closed his speech delivered on the occasion in the following words:

> We did not expect to conquer the whites. They have too
> many houses, too many men. I took up the hatchet, for my

part, to revenge injuries which my people would no longer endure. Had I borne them longer without striking, my people would have said: 'Black Hawk is a woman; he is too old to be a chief; he is no Sac.' These reflections caused me to raise the war whoop. I say no more. It is known to you. Keokuk once was here; you took him by the hand, and when he wished to return to his home you were willing. Black Hawk expects like Keokuk, he shall be permitted to return, too.

By order of the president, Black Hawk and his companions who were in confinement at Fortress Monroe, were set free on the 4th day of June, 1833. After their release from prison they were conducted in charge of Major Garland through some of the principal cities that they might witness the power of the United States and learn their own inability to cope with them in war. Great multitudes flocked to see them wherever they were taken and the attention paid them rendered their progress through the country a triumphal procession instead of prisoners transported by an officer. At Rock Island the prisoners were given their liberty amid great and impressive ceremony. In 1838 Black Hawk built him a dwelling near Des Moines, this state, and furnished it after the manner of the whites and engaged in agricultural pursuits, together with hunting and fishing. Here, with his wife, to whom he was greatly attached, he passed the few remaining days of his life. To his credit it may be said that Black Hawk remained true to his wife and served her with devotion uncommon among Indians, living with her upwards of forty years.

At all times when Black Hawk visited the whites he was received with marked attention. He was an honored guest of the Old Settlers' reunion in Lee county, Illinois, and received marked tokens of esteem. In September, 1838, while on his way to Rock Island to receive his annuity from the Government, he contracted a severe cold, which resulted in an intense attack of bilious fever, and terminated his life October 3rd.

After his death he was dressed in the uniform presented him by the president while in Washington. He was buried in a grave six feet in depth, situated upon a beautiful eminence. The body was placed in the middle of the grave, in a sitting position upon a seat constructed for the occasion. On his left side the cane given him by Henry Clay was placed upright, with his right hand resting upon it. His remains were afterward stolen and carried away but they were recovered by the governor of Iowa and placed in the museum in Burlington, of the Historical Society, where they were finally destroyed by fire.

The territory known as the "Black Hawk Purchase," although not the first portion of Iowa ceded to the United States by the Sacs and Foxes, was the first opened to actual settlement by the tide of emigration which flowed across the Mississippi as soon as the Indian tide was extinguished. The treaty, which provided for this session, was made at a council held on the west bank of the Mississippi where now stands the city of Davenport, on ground now occupied by the Chicago, Rock Island & Pacific Railroad Company, September 2, 1832. This was just after the Black Hawk war and the defeated savages had retired from east of the Mississippi.

At the council the Government was represented by Gen. Winfield Scott and Governor Reynolds, of Illinois, Keokuk, Pash-a-popo, and some thirty other chiefs and warriors were there. By this treaty the Sacs and Foxes ceded to the United States a strip of land on the eastern border of Iowa, fifty miles wide, from the northern boundary Missouri to the mouth of the Upper Iowa river, containing about 6,000,000 acres. The western line of the purchase was parallel with the Mississippi.

In consideration for this cession, the United States agreed to pay annually to the confederated tribes, for thirty consecutive years, $20,000 in specie, and to pay the debts of the Indians at Rock Island, which had been accumulating for seventeen years and amounted to $50,000, due to Davenport & Farnham, Indian traders. The Government also donated to the Sac and Fox women and children, whose husbands and fathers had fallen in the Black Hawk war, thirty-five beef cattle, twelve bushels of salt, thirty barrels of pork, fifty barrels of flour and six thousand bushels of corn.

The treaty was ratified February 13, 1833, and took effect on the 1st of June following, when the Indians quietly removed from the ceded territory and this fertile and beautiful region was opened by white settlers.

By the terms of the treaty, out of the "Black Hawk Purchase" was reserved for the Sacs and Foxes four hundred square miles of land, situated on the Iowa river, and including within its limits Keokuk village, on the right bank of that river. This tract was known as Keokuk's reserve and was occupied by the Indians until 1836, when, by a treaty made in September between them and Governor Dodge, of Wisconsin territory, it was ceded to the United

States. The council was held on the banks of the Mississippi above Davenport, and was the largest assemblage of the kind ever held by the Sacs and Foxes to treat for the sale of land. About one thousand of their chiefs and braves were present, Keokuk being the leading spirit of the occasion and their principal speaker.

By the terms of this treaty the Sacs and Foxes were removed to another reservation on the Des Moines river, where an agency was established at what is now the town of Agency, in Wapello county. The Government also gave out of the "Black Hawk Purchase," to Antoine LeClaire, interpreter, in fee simple, one section of land opposite Rock Island and another at the head of the first rapids above the island, on the Iowa side. This was the first land title granted by the United States to an individual in Iowa.

. . .

In May, 1843, most of the Indians were removed up the Des Moines river, above the temporary line of Red Rock, having ceded the remnants of their land in Iowa to the United States, September 21, 1837, and October 11, 1842. By the terms of the latter treaty, they held possession of the "New Purchase," until the autumn of 1845, when most of them were removed to their reservation in Kansas, the balance being removed in 1846.

. . .

After the "Black Hawk Purchase" immigration to Iowa was rapid and steady, and provision for civil government became a necessity. Accordingly, in 1834, all the territory comprising the present States of Iowa, Wisconsin and Minnesota, was made subject to the jurisdiction of Michigan territory. Up to this time there had been no county or other organization in what is now the State of Iowa ... In September of 1834, therefore, the Territorial Legislature of Michigan created two counties on the west side of the Mississippi river — Dubuque and Des Moines — separated by a line drawn westward from the foot of Rock Island."

Iowa would not have become a state yet when Thomas and his family arrived and since we don't know an exact date, only the range, according to Getchell, from 1836 to 1840, it may not have even yet been Iowa Territory. Iowa Territory was split off from Wisconsin Territory in 1838. Getchell states that Thomas entered on two hundred acres of virgin public domain or government land, and as pioneers on that land they set about the arduous task of breaking and developing the land.

Research by cousin Max Fitz revealed that Thomas first entered on 80 acres of government land in 1843 and then added another 40 acres in each year, 1850 and 1851, and another 80 acres in 1852. In 1867 he acquired 40 acres from Chris Neiderhiser which brought his land holdings to 280 acres. Their land would have been just south of what is now Cedar Rapids and was in the vicinity of the town of Ely. We know there is quality farmland in that area of excellent fertility. The area they chose would probably have been gently rolling with tall grass, perhaps with some trees since it would have been only a few miles from the Cedar River. Grassland, which is probably what they chose, would not have been as difficult to break as woodlands with stumps to contend with, but soil that had never seen a plow would have been arduous to break. Again thanks to cousin Max Fitz, we have maps, included at the end of this chapter, of exactly what land was entered by Thomas and Charlotte.

Our great-great grandparents, Thomas and Charlotte Fitz were the true pioneers of the Fitz family that emigrated to Iowa. They, however, were followed by two of Thomas's brothers, Wiley Fitz who entered on land in Linn County in 1845 and Henry Fitz who entered on land also in Linn County in 1853. Henry later moved on to Greene County but both Thomas and Wiley we believe lived their entire lives in Linn County.

While numerous Indian tribes called what would become Iowa Territory home as the 19th century began, by 1845 nearly all Indians had left the area, the last being the Sioux in 1852. In the treaty between the United States and

Thomas & Charlotte Fitz

the Sac and Fox tribes signed in 1842 and often called the New Purchase of 1842, the Indians agreed to sell the land in Iowa west of the Mississippi River and north of the Missouri line and to relocate in Kansas. By 1856 the Meskwaki (Fox) tribe had convinced Iowa Governor James Grimes to allow them to purchase land in Iowa and so the Meskwaki Settlement in Tama County began which still exists today

The Fitz genealogy by Sylvia Fitts Getchell lists two of Thomas and Charlotte's children that did not survive into adulthood: Mary D. who was born on February 18, 1840 but died at childbirth and William who was born on April 13, 1849 but only lived one month and 2 days.

In addition to the two children, Elizabeth Ann and John Abraham, who had been born in Ohio before the family moved westward to Iowa and in addition to the two who did not survive into adulthood, Thomas and Charlotte were parents to Rebecca Jane born December 1, 1837, Henry Harrison born August 31, 1842, Martha Malinda born November 6, 1844, David born in 1847, Washington Wiley born September 6, 1850 and the baby of the family Alvin Mindon, our Grandpa Fitz, born February 11, 1855.

Thanks to Cousin Carrie (Fitz) Lambert there are some stories that need to be retold. The first one occurred during the Civil War (1861–1865). I, Clarence, one of the authors, remember our Dad, Ralph, talking about a Fitz relative who endured the horrendous treatment of prisoners at Andersonville Prison during the Civil War. I had no idea though that that relative was Grandpa Fitz's brother, John Abraham. Cousin Carrie quotes from John Abraham Fitz's obituary:

> He enlisted in April 1861 at Mt. Vernon, Iowa in Company A of the 13th Infantry and also belonged to Crocker's Brigade. He served all four years of the War and took an active part in seventeen heavy battles and in numerous small ones, was in Sherman's famous March to the Sea and was taken prisoner at Atlanta, Georgia, and in Andersonville Prison 4 months and in Libby Prison 3 months until he was released at the close of the Civil War.

Forty-five thousand Union soldiers were imprisoned at Andersonville and during the fourteen months the prison existed thirteen thousand died of disease, poor sanitation, malnutrition, overcrowding and exposure.

One of the prisoners described the sanitation issue,

> The camp was covered with vermin all over. You could not sit down anywhere. You might go and pick the lice all off of you, and sit down for a half a moment and get up and you would be covered with them. In between these two hills it was very swampy, all black mud, and where the filth was emptied it was all alive, there was a regular buzz there all the time, and it was covered with large white maggots.

That was Andersonville in Georgia. John Abraham also spent three months in Libby Prison near Richmond, Virginia where we think he was moved after his four months at Andersonville. Conditions at Libby were clearly better than Andersonville but still plagued with overcrowding and lack of food.

Another Fitz of renown was Dr. Reginald Heber Fitz whom I (Clarence) first became aware of because of a historical display at the Mayo Clinic in Rochester, Minnesota. As a physician he was the first to understand and describe the condition known as appendicitis. Others would then develop the surgery, appendectomy, as a treatment for the disease, a treatment that saved the life of our father Ralph in the early 1950's. Reginald Fitz was born May 5, 1843 in Chelsea, Massachusetts and died September 30, 1913. He received his M.D. from Harvard in 1909, studied medicine in Vienna, Berlin and Paris, then came back to the United States and became an associate professor of medicine at Harvard for 14 years and then assistant dean of the Harvard School of Medicine. At various times he served on the staffs of several hospitals including the Mayo Clinic.

Cousin Carrie in her genealogy paper writes about a Henry Fitz who was born in Newburyport, Massachusetts in 1808 who was a telescope maker who attracted the attention of the world. Quoting from Cousin Carrie's paper, "In 1835 he constructed his first Reflecting Telescope, attracting the attention and winning the patronage of many astronomers throughout the world. His methods were of his own invention and his business became so successful that he moved to a larger plant in New York City in 1845. He made use of local polishing, 15 years before the process was developed in Europe. His instruments were used in the U.S. Astronomical Expedition to the Southern Hemisphere ... several comets were discovered with his precise instruments. ... His ambition was to build a 24-inch telescope and he was about to sail for Europe to select the glass for it when he died at 55 years of age in 1863."

Thus began the Fitz family in Iowa that includes the authors of this book and beyond. From this point on this book deals with people with whom the authors were either personally acquainted and often related to or personally knowledgeable about. We have shown, we think, great doubt about being related to John Fitz the bencher. In trying to search for the real Robert, we came to a dead end.

Following the line backwards we come to Henry Fitz Randolph who apparently dropped the Randolph from his name and became Henry Fitz. Not an uncommon practice in those days. Does it matter? Probably not unless you are a genealogist or a romantic who was dead set on proving that you descended from the royalty of England or you wanted to trace your roots back to Jamestown. We have shown, we think, that the Fitz family name came to America with the first permanent colony in America at Jamestown and followed shortly after with the Pilgrims in New England. Maybe not our Fitz family branch but still a Fitz family. And we have shown, we think, that the Fitz family helped to push back the frontier and settle the west at least all the way to Iowa. Obviously, there is more painstaking research to be done in order to fit the pieces together. The four pages at the end of this chapter show the location and property records of the Thomas B. Fitz land holdings and others of his family.

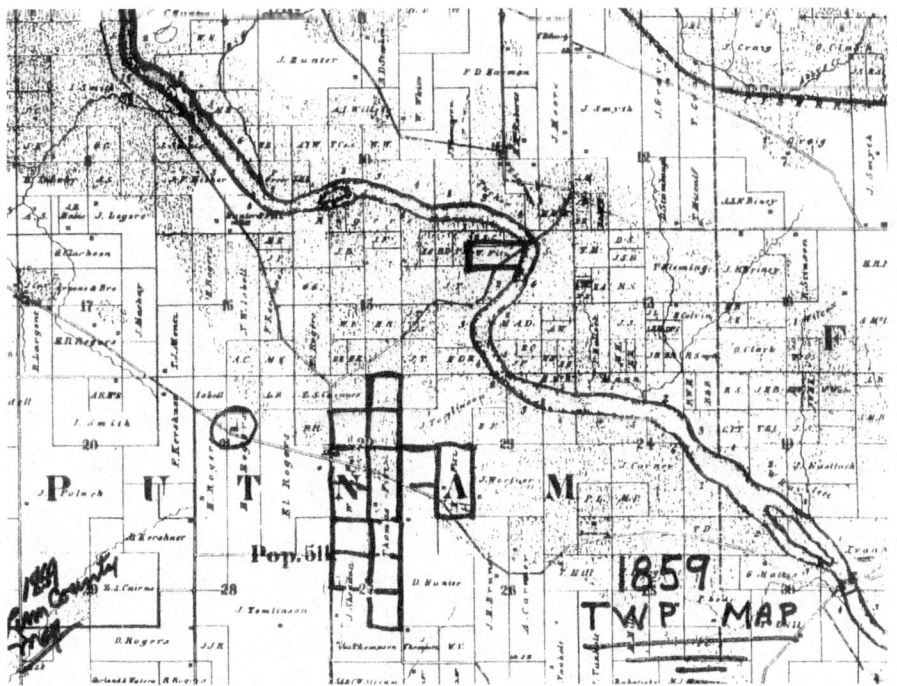

Thomas B. Fitz land denoted by bold lines.

Rogers Grove Cemetery, Ely IA

```
L
I  BURIAL INFORMATION:
N  ...Tombstone...                               Lcn
E  Surname   Given      Birth     Death     Bk-Lot-No Owner  .......Remarks...............................................
- - - - - - - - - - - - - - - - - - - - - - - - - - - - - - - - - - - - - - - - - - - - - - - - - - - - - - - - - - - - - - -
 1 Bates, Louisa"? Maybe LOIS    1860-09-07 06-04-03 Fitz,  LooseStone"J.H.Bruner"S/B 6-3-1/2/3? (#24pg411);age 1y6m?dy.
 2 Fitz                           1875-09-06          d/o John Abraham & Caroline(Upmeyer) Fitz (pg411).
 3 Fitz      Can't read: 2 stones?           05-13-05 Fitz,  [ ? Charlotte (Bruner) Fitz ? w/o Thos(3).
 4 Fitz                                       06-04-01 Fitz,  [
 5 Fitz      Can't read "Emma ?"             05-13-04 Fitz,  [
 6 Fitz                                       10-05-01 Fitz,  [
 7 Fitz      Jane       1806-10-13 1867-12-30 10-05-03 Fitz,  w/o Henry Fitz(3)(pg404);Jane Lindsey(Linsy?);age 61y2m7dy.
 8 Fitz      Thomas     1810-04-01 1875-02-07 05-13-03 Fitz,  s/o John(2); age 64yr10mo7day.
 9 Fitz      Rebecca J. 1837-12-01 1846-12-10 05-13-01 Fitz,  d/o Thos & Charlotte Fitz.
10 Fitz      John H.    1844       1846-03-08 10-05-02 Fitz,  s/o Wiley & Sarah(Rogers) Fitz; age 2y?m?d.
11 Fitz      William    1849-04-13 1849-05-15 05-13-02 Fitz,  s/o Thos(3) & Charlotte(Bruner) Fitz
12 Fitz      Mary Sophia 1852-09-19 1875-02-26 09-09-06 Neider d/o John & Jane Neiderheiser, m.David Fitz(4)(pg405).
13 Fitz      Inf. dau.  1877-07-30 1877-07-30 06-04-02 Fitz,  d/o Wash.Wiley & (Barb.)Ellen(Blessing) Fitz(pg412).
14 Kershner  Ferdinand  1821-05-13 1916-05-11 08-05-04 Kershn h/o Elizabeth (Rogers)
15 Kershner  Elizabeth(Ro 1826-12-27 1893-01-09 08-05-05 Kershn d/o Daniel C. & Elizabeth (Fitz) Rogers
16 LeGore    John       1816-04-18 1901-07-24 05-06-05 LeGore p/o Anne Eliz. Fitz(James Wiley pg410).
17 LeGore    Ellen E.   1821-12-23 1878-05-22 05-06-04 Muntz  p/o Anne Eliz. Fitz(James Wiley pg410).
18 Neiderhe  Jane       1823-07-25 1894-06-13 09-09-02 Nederh p/o Mary Sophia Fitz
19 Neiderhe  John       1824-03-15 1897-11-21 09-09-01 Nederh p/o Mary Sophia Fitz
20 Rogers    Robt Rodman 1776      1853-07-17 07-04-04 Rogers s/o Thos & Ann (Rodman) Rogers; h/o Mary (Fitz) (pg400).
21 Rogers    Mary (Fitz) 1784      1866-12-07 07-04-03 Rogers d/o Henry & Jane [ ? ) Fitz (pg400); w/o Rbt Rodman Rogers.
22 Rogers    Daniel C.  1789-04-22 1854-05-22 08-04-06        s/o Thos & Ann(Rodman) Rogers.
23 Rogers    Elizabeth  1794       1874-06-18 08-04-05        d/o Henry Fitz(1); sis of John(2).
24
25
26
27         -----------------------------   -----------------------------
28         Blk 6                      ;    Blk 7
29         ;        #1,13             ;    ;              #20,21          ;
30         -----------------------------   -----------------------------
31         -----------------------------   -----------------------------
32         Blk 5     #9,11:1stone     ;    Blk 8         #14,15          ;
33         ;           #8             ;    ;             #22,23          ;
34         -----------------------------   -----------------------------
35         -----------------------------   -----------------------------
36         Blk 4                      ;    Blk 9
37         ;                          ;    ; #12,18,19                   ;
38         -----------------------------   -----------------------------
39         -----------------------------   -----------------------------
40         Blk 3                      ;    Blk 10
41         ;                          ;    ;              #7,10          ;
42         -----------------------------   -----------------------------
43         -----------------------------            --------------------
44         Blk 2                      ;    ;    ; Blk 11                 ;
45         ;                          ;    ;    ;                        ;
46         -----------------------------   ; Church; --------------------
47         -----------------------------   ;    ; --------------------
48         Blk 1                      ;    ;    ; Blk 12                 ;
49         ;                          ;    ;    ;                        ;
50         -----------------------------   ;----; --------------------
```

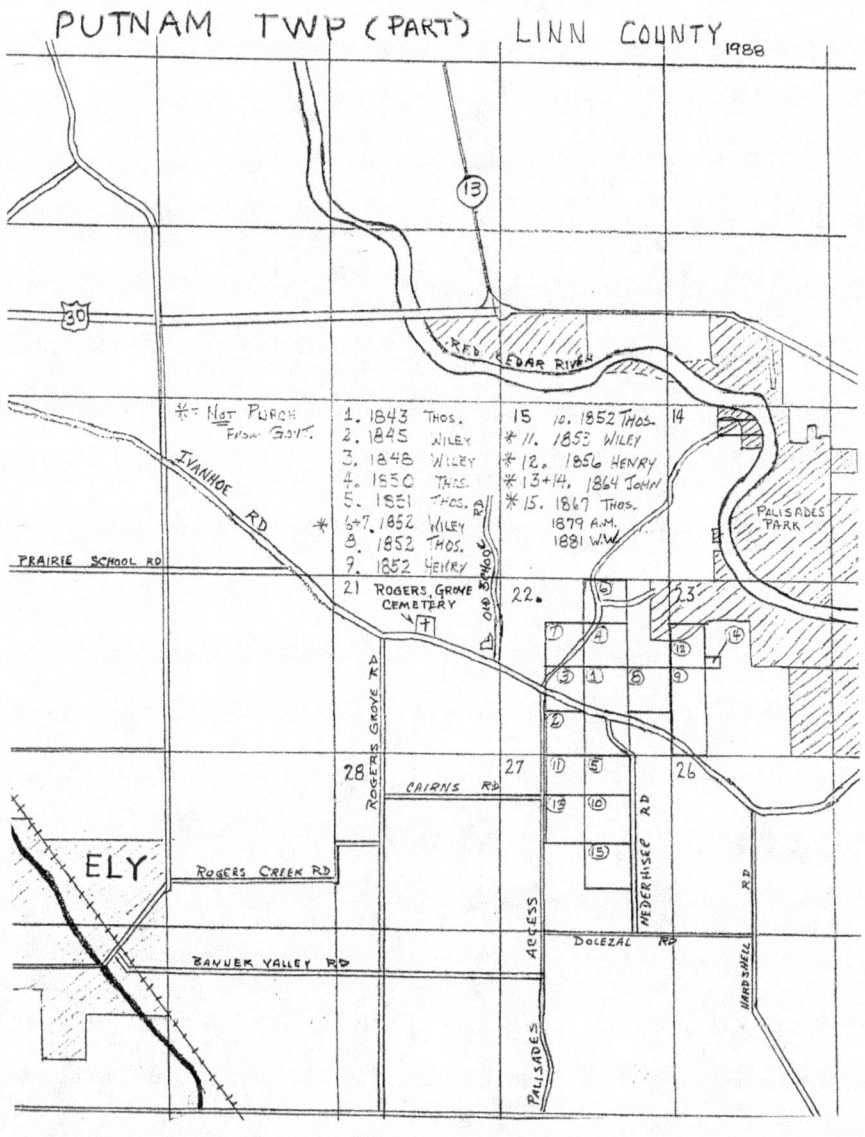

PUTNAM TWP (PART) LINN COUNTY 1988

RED CEDAR RIVER

PALISADES PARK

✳ = NOT PURCH FROM GOV'T.

1. 1843 THOS.
2. 1845 WILEY
3. 1848 WILEY
4. 1850 THOS.
5. 1851 THOS.
6+7. 1852 WILEY
8. 1852 THOS.
9. 1852 HENRY

15
✳ 10. 1852 THOS.
✳ 11. 1853 WILEY
✳ 12. 1856 HENRY
✳ 13+14. 1864 JOHN
✳ 15. 1867 THOS.
    1879 A.M.
    1881 W.W.

14

21 ROGERS GROVE CEMETERY

22.

23

28

27

26

CAIRNS RD

ELY

ROGERS CREEK RD

ROGERS GROVE RD

OLD SCHOOL RD

NEDERHISER RD

PALISADES ACCESS

DOLEZAL RD

HARDSHELL RD

BANNER VALLEY RD

IVANHOE RD

PRAIRIE SCHOOL RD

30

13

```
L   LINN CO., Putnam Twp: Fitz land record
I
N   New                                              Sec-
E   Owner        Source          Date      Legal    Twp-Age Other      Tfr to:
---------------------------------------------------------------------------------------
 1 Fitz, Thomas  Govt            1843/09/15  W 1/2 SE1/4 22-82-6 80A
 2 Fitz, Wiley   Govt            1845/05/16  SE1/4 SW1/4 22-82-6 40A
 3 Fitz, Wiley   Govt            1848/07/07  NE1/4 SW1/4 22-82-6 40A
 4 Fitz, Thomas  Govt            1850/03/14  SW1/4 NE1/4 22-82-6 40A
 5 Fitz, Thomas  Govt            1851/02/04  NW1/4 NE1/4 27-82-6 40A
 6 Fitz, Wiley   Berryhill, Chas & 1852/01/16 NW1/4 NE1/4 22-82-6 40A     Neiderhiser, John 1876/
 7 Fitz, Wiley   Berryhill, Chas  1852/01/20  SE1/4 NW1/4 22-82-6 40A     Neiderhiser, John 1876/
 8 Fitz, Thomas  Govt            1852/05/12  SE1/4 SE1/4 22-82-6 40A
 9 Fitz, Henry   Govt            1852/06/24  W 1/2 SW1/4 23-82-6 80A
10 Fitz, Thomas  Govt            1852/08/12  SW1/4 NE1/4 27-82-6 40A
11 Fitz, Wiley   Berryhill, Chas  1853/08/20  NE1/4 NW1/4 27-82-6 40A
12 Fitz, Henry   McClelland, John 1856/03/01  SE1/4 NW1/4 23-82-6 40A
13 Fitz, John A. Neiderhiser, Chris 1864/04/07 SE1/4 NW1/4 27-82-6 40A, Less 1 A. for MethChurch
14 Fitz, John A. Neiderhiser, Chris 1864/04/07 SE1/4 NW1/4 23-82-6 ? 5A
15 Fitz, Thomas  Neiderhiser, Chris 1867/01/11 NW1/4 SE1/4 27-82-6 40A
16 Fitz, Geo. S. Mtge            1873/02/15                    Grantor: James Moreland
17 Fitz, W. W.   Fitz, Thos heirs 1875/12/23  W 1/2 NE1/4 22-82-6 80A
18 Fitz, Wash. W. Fitz, Charolette 1876/02/18  W 1/2 NE1/4 27-82-6 80A  Wife Ella
19 Fitz, W. W.   Fitz, Thos heirs 1876/12/23  NW1/4 SE1/4 27-82-6 40A
20 Fitz, Alvin M. Fitz, John A.    1879/02/24  NW1/4 SE1/4 27-82-6 40A    Fitz, W. W. 1881/02/17
21 Fitz, W. W.   Fitz, Alvin Min  1881/02/17  NW1/4 SE1/4 27-82-6 40A
22
23
24
25
26
27
```

Henry H. Fitz

James Wiley Fitz

John Fitz

CHAPTER 4

GRANDPA FITZ

Alvin Mindon Fitz

A lvin Mindon Fitz, the youngest of Thomas and Charlotte Fitz's children, was born February 11, 1855 in Ely, Linn County, Iowa. Property records indicate that he owned 40 acres in Linn County just south of one of his Dad's properties which he acquired in 1879 from John A. Fitz whom we are assuming was his brother John Abraham. In 1881 he transferred that property to W.W. Fitz who we think was probably another brother, Washington Wiley Fitz. (Please see the last four pages of Chapter 3)

Alvin Mindon Fitz

That same year, when he was 26 years old (1881), he followed in his father's footsteps, struck out on his own, moved to Greene County, Iowa, and purchased 160 acres of virgin prairie (section 29 of Cedar township). He married Adeliza (Addie) Eliza Ritchie on March 9, 1882 and together they transformed 160 acres of virgin prairie into what Greene County history describes as "one of the pleas-antest homes and most productive farms in that community." Grandpa Fitz farmed

Adeliza (Addie) Eliza Ritchie Fitz

until 1915 when he moved to town, presumably Rockwell City, and later to Hampton.

Great Aunt Martha Malinda was eleven years older than Grandpa Alvin Mindon. Her future husband, John Henry Hunter enlisted in the Civil War Union Army in 1862 at age 19 and was wounded in action in 1863 after only nine months of service. He was mustered out of the army in 1865 and married Great Aunt Martha Malinda in 1867. Great Uncle John Henry Hunter and Great Aunt Martha Malinda were the first of the Fitz's we believe to farm in Greene County near Churdan, Iowa, where in 1866 Great Uncle John Henry broke an acreage of prairie with a team of oxen which he and his future wife farmed until 1882. In 1882 Great Uncle John Henry rented his land and he and a partner, Livingston, built one of the first two buildings in Churdan and started a general merchandise store. Great Aunt Martha Malinda died at age 90 after a broken hip resulted in pneumonia.

Great Uncle Henry Harrison Fitz, we think, was the next of the family to be a farmer in Greene County. Like his brother-in-law John Henry Hunter and his older brother John Abraham, he had enlisted in the Civil War Union Army at age 19 and he too was discharged in 1865. The 1870 census tells us that he was a farmer in Greene County but in Kendrick Township and apparently Great Uncle David was living with him at the time of the census. We have very little information on Great Uncle David but we know that his wife Mary died in 1875 at the age of 22 and that Great Uncle David was again counted along with Great Uncle Henry Harrison's family in the 1880 census. The 1880 census though indicates that Great Uncle Henry Harrison was now a farmer in Cedar Township. We have found no evidence that he owned a farm in Greene County. In 1887 he moved to Ness City, Kansas but about four years later moved on to Northwest Oklahoma Territory where he homesteaded two miles east of Grand. There they lived on his homestead until his death about eight years later in 1909, at age 67.

Grandpa Alvin Mindon was the next to arrive in Greene County. In 1881 at age 26 he purchased a quarter section of prairie in Cedar Township near Churdan. Deeds were written in long hand at that time but what follows is an excerpt from that deed copied as accurately as we could considering the handwriting.

This indenture made this eleventh day of April in the year of our lord, one thousand eight hundred and eighty-one between Orrison Wheeler of Three Rivers, St Joseph County, Michigan, of the first part and Alvin M. Fitz of Greene County, State of Iowa, of the second part, Witnesseth that the said party of the first part for and in consideration of the sum of Sixteen Hundred Dollars to her in hand paid by the said party of the second part, the receipt whereof is hereby confessed and acknowledged, have Granted, Bargained, Sold, recuiesed (?), Released, Aliened and confirmed, and by these presents do Grant, bargained, Sell, reuiese(?), release, alien and confirm unto the said party of the second part his heirs and assigns forever.

The following described land to wit: The Southwest quarter of Section No. Twenty Nine (29) In Township No. Eighty five (85) North Range No. Thirty Two (32) West of the 5th principal meridian containing according to the original survey, be the same more or less. Situated in Greene County, State of Iowa.

The sellers name, who is identified by the notary as being a widow, is spelled at least two different ways, maybe three: Orrison and Orissa. We prefer to think the sellers name was Orissa Wheeler.

Great Uncle Washington Wiley was the last of Thomas and Charlotte's children to establish a farm in Greene County. In the1880 census Grandpa Alvin Mindon was counted along with his brother Washington Wiley's family in Linn County but soon after we would assume he must have started the process of buying land in Greene County. Great Uncle Washington Wiley was listed as a farmer on the 1880 census and in 1884 he purchased 160 acres of raw prairie in Greene County which he developed into a fine productive farm. He raised livestock and specialized in Poland China hogs which he shipped to city markets. After 1902 he rented his land and spent his time raising squabs which he shipped to Chicago.

Grandpa Alvin Mindon and Grandma Addie sold their property in Greene County on February 17, 1901 to J.W. Wynkoop of Guthrie County for six thousand four hundred dollars. By then deeds had become much less flowery and so the property description on the deed reads: "The south west quarter of Section twenty-nine (29) Township Eighty Five (85) North Range Thirty Two (32) west 5th P.M. Iowa containing 160 acres more or less according to government survey thereof…" They had transformed the raw prairie into a productive farm and were ready for the next challenge. All of their children had been born on this farm except for the youngest, Aunt Hazel.

On February 28, 1901, Grandpa Alvin Mindon purchased from Mary L. Spitzbarth and her husband, J.N. Spitzbarth, for the price of $7,496.25, "The northeast fractional quarter of Section Five (5) in Township Eighty-seven (87) North, in Range thirty-two (32) West of the 5th P.M. Iowa. Containing 187.43 acres according to Government survey, except legal highways." This property was located in Calhoun county and was owned by Grandpa Alvin Mindon until his death on January 8, 1944. We have no information on whether it was raw prairie or whether it had been broken.

Following Grandpa Alvin Mindon's death, and according to his will, the Calhoun County farm was titled to his children, C.C. Fitz, Myrtle Fitz, Earl Fitz, Ralph Fitz, Ray Fitz, Velma Fitz Heideman and Hazel Marken on February 13, 1946.

On March 1, 1946, C.C. Fitz and his wife Helen, Myrtle Fitz, Ralph Fitz and his wife Edith, Ray Fitz and his wife Mildred, all of Franklin County, Iowa; Earl Fitz of Fayette County, Iowa; and Hazel Fitz Marken and her husband Ralph, of Black Hawk County, Iowa, all sold their interest in the farm they had inherited to Velma Fitz Heideman and her husband Hiram H. Heideman. The farm was now described as "The Northwest Fractional Quarter (NE frl ¼) of Section 5, Township 87 North, Range 32, west of the 5th P.M., Calhoun County, Iowa, except the Railway right-of-way" … "containing 180.70 acres more or less"

On the same day, March 1, 1946 Velma Fitz Heideman and her husband Hiram H. Heideman sold a one-half interest in the farm to Henry E. Heideman and his wife Lilian Heideman of Cook County, Illinois.

On January 27, 1947, Velma Fitz Hiedeman and her husband Hiram H. Heideman, apparently having secured a loan from the Federal Bank of Omaha, repurchased the individual half interest in the farm from Henry E. Heideman and his wife Lilian Heideman.

Upon the death of Hiram H. Heideman and according to his will, the title of the farm passed to his wife Velma Fitz Heideman and was then described as:

> An undivided 15/16th interest in and to the Northeast Fractional Quarter of Section 5 in Township 87 North, Range 32 West of the 5th P.M., Calhoun County, Iowa, including the railroad right-of-way conveyed to them by Quit Claim Deed recorded in Book 101, Page 347 of the Records of the Calhoun County, Iowa Recorder.

On March 1, 1976 Velma Fitz Heideman sold the farm to Dean Heideman and his wife Betty Heideman and V.C. Heideman and his wife Muriel Heideman. The farm was at that time described as containing "approximately 90.34 acres more or less."

On September 13, 1988, Velma Fitz Heideman deeded the farm with love and affection to Dean Heideman and V.C. Heideman. At this occurrence the farm was legally described as "The Northwest Quarter of the Northeast Fractional Quarter (NW 1/4 NE Frl ¼) and the Southwest Quarter of the Northeast Fractional Quarter (SW ¼ NE Frl ¼), All in Section 5 (5), Township Eighty-seven (87) North, Range Thirty-Two (32) West of the 5th P.M., containing 103.88 acres, more or less."

At the 2001 Iowa State Fair, the farm that Grandpa Alvin Mindon had purchased in 1901 was recognized as a century farm, meaning it had been in the same family for 100 years. In reporting on that occasion, the *Farm News/ Fort Dodge, Iowa* interviewed Dean Heideman. Here are excerpts from that interview. This interview was conducted by Thayne Cozart, a *Farm News* Correspondent.

> Standing on an earthen mound on a quiet May morning near the west border of the family's 190-acre century farm southeast of Rockwell City in Calhoun County, Dean Heideman muses, "That's where the old Fort Dodge, Des Moines and Southern Inter Urban electric train from Des Moines ran early last century. It's kind of hard to believe that once hundreds of persons traveled right through here every day conducting the activities of their lives."

Members of the Heideman and Fitz families have likewise conducted portions of their lives for 100 years at this scenic location.

Heideman continues,

> "Alvin Fitz was the original family member who purchased the farm in 1901. I remember him telling of helping build the track bed for the electric train using his team of horses and a slip. And I can recall my dad rode the Inter Urban into town during snowstorms."

Then pointing to the nearby abandoned gravel pit, he recalls,

> "As a youngster I spent many an enjoyable day playing in the gravel pit. We called it 'Wolf Canyon.' And our children and their cousins enjoyed it as much as the children of my generation did. Sometimes, we camped overnight or fished in the creek.
>
> "And this spot must have been popular for the Indians, too, because we've found numerous Indian arrowheads and relics close by," he says.
>
> "Alvin Fitz and his brothers [we believe he meant to say son's instead of brothers — we think his son, Uncle Earl farmed the farm up until 1946 when Aunt Velma and Uncle Hiram bought the farm from Grandpa Alvin's heirs] operated the farm following Alvin's purchase right up to 1944 or 1945. That's when Hiram and Velma Heideman bought the farm from the estate. In 1990, the current owners inherited the balance of the farm that they hadn't earlier purchased.
>
> "The history of the farmstead is interesting. The first building was a barn, which was later converted to a corn crib with a new barn. The next building on the farmstead was the house — an elegant two-story, five-bedroom home with 10-feet six-inch ceilings and double French doors with transoms for ventilation.
>
> "It was an elegant old house, Betty Heideman comments.

Other outbuildings followed. The second barn was built in 1908. Much of the cement and brick for construction of the barn and garage was from materials excavated from the gravel pit.

As of 2001, the farmstead stood vacant.

Grandpa Alvin Mindon Fitz retired and moved to Rockwell City, Iowa in 1915 and sometime later moved to Hampton, Iowa where he lived when he died January 8, 1944.

**1902
Alvin Mindon Fitz Farm
Rockwell City, IA**

**Later, this becomes
Velma & Hiram
Heideman's Farm**

Alvin Mindon Fitz farmhouse

Alvin Mindon Fitz
Back View

Alvin Mindon Fitz home in Hampton, Iowa, front and rear views. Who is looking out the upstairs window?

Grandma Addie's tombstone

Grandpa Alvin's tombstone

Clarence remembers Grandpa Fitz

Grandpa Fitz (Alvin Mindon Fitz) and Aunt Myrtle lived together in Hampton. My memories of Grandpa Fitz are pretty meager. I remember that he had an area in the basement of his house in Hampton that was his. His wooden rocking chair sat in front of the big furnace. I remember rugs or blankets around the chair that made it a warm cozy area. I believe that Grandpa Fitz smoked a pipe but that area is the only place I ever saw him smoke it.

I remember having dinner (lunch) at Grandpa's house. It was fun watching Aunt Myrtle raise the dumbwaiter from the basement to the kitchen. I remember that the dumbwaiter was the storage place for the bowl of butter and the inevitable beet pickles that seemed to be served at every meal.

Grandpa Fitz came to stay at the farm for a couple of days. I remember that Grandpa gathered the eggs from the chicken house for my mother, but as he carried the basket of eggs to the basement he fell. The broken eggs were quite a mess but of more concern was the cut on Grandpa's forehead. I remember going with daddy to take Grandpa to see Dr. Martin in Latimer. Dr. Martin thought Grandpa's forehead needed some stitches but Grandpa was pretty old so he decided to not use any anesthetic.

Grandpa Fitz was in the hospital I heard my folks say — pneumonia I think. The next time we went to Grandpa's house was because Grandpa had died. The house smelled funny, different than I was used to, like fresh paint. Apparently, someone had done some painting while Grandpa was in the hospital (maybe my parents did that). That would be a good time to do it before Grandpa came home from the hospital. For years after that, whenever I smelled that particular smell of fresh paint, I thought of Grandpa not coming home from the hospital.

The only other remembrance I have is of the auction sale to dispose of Grandpa's household belongings. As a young boy, I took it upon myself to be the pretend clerk of the sale, writing down each item and the price that it sold for. How proud I was when the actual auctioneer and clerk asked me what price I had recorded for a certain item that their records were unclear about.

Alvin Mindon Fitz family

Grandma Addies family. Back row: Ellie, Pearl, Will, Ann, Addie. Front Row:
Great Grandma Ritchie, Great Grandpa Ritchie.

Alvin Mindon Fitz family. Left to right: Myrtle, Charlie, Velma, Grandma Addie, Aunt Mary Ritchie Porter, Hazel, Grandpa Alvin Mindon, Olae, Ralph, Earl, Ray & Lloyd

Alvin Mindon Fitz family (left). Back row: Charlie, Myrtle, Earl, Olae. Front row: Lloyd, Ray, Rulph

Alvin Mindon holding Clarence.

Alvin Mindon Fitz with ?

Alvin Mindon Fitz family, 1938.

1938 gathering at Grandpa Alvin Mindon's

* * * * * * *

No. 2.

J.L. Sinclair and wife Mabel C.   WARRANTY DEED

        to   Dated Feb. 23, 1910
                                              Filed Mar. 16, 1910
Addie E. Fitz   Book "47" page 127

      Conveys the SE¼ 28-92-20 except 1 acre tract in the SE Corner
thereof.

* * * * * * *

No. 3.

IN THE DISTRICT COURT OF IOWA, IN AND FOR FRANKLIN COUNTY

In the Matter of the Estate   IN PROBATE

        of   Cause No. 2163

Addie E. Fitz, deceased

      Addie E. Fitz died Sept. 20, 1925, testate, a resident of
Franklin County, Iowa, survived by A.M. Fitz, her spouse. After Notice
her will which was admitted to probate on Nov. 4, 1925 provides: I give
to my beloved husband, A.M. Fitz, one-third of all the real and personal
property of which I may die seized. Subject to above bequest I give to
my living children, with the exception of Loyd J. Fitz, all the rest
and residue of my real and personal property, share and share alike.
Probate Inventory lists as beneficiaries, C.C. Fitz, Myrtle Fitz, Earl
Fitz, Lloyd J. Fitz, Ralph Fitz, Ray Fitz, Velma Fitz, Hazel Marken her
children all of whom were of legal age and A.M. Fitz, as surviving husband
and lists under real estate the SE¼ of Section 28-92-22. Final report
approved and estate closed Aug. 6, 1927.

* * * * * * *

FRANKLIN COUNTY ABSTRACT CO.

Hampton, Iowa

Page 1.

Grandma Addie's will.

IOWA, JANUARY 13, 1944

Hold Final Rites
For Alvin M. Fitz

Hampton Resident Dies Of Bronchial Pneumonia

Funeral services for Alvin M. Fitz, age eighty-eight, longtime resident of Hampton, who died last Saturday, January 8th, after an illness of bronchial pneumonia, were held last Tuesday afternoon in the Methodist church, in Hampton, with Rev. W. C. Cleworth in charge. Burial was made in the Hampton cemetery. Mr. Fitz had been in failing health for several years.

Mr. Fitz was born near Cedar Rapids, on February 11, 1855. When a young man he went to Greene county, where he was married to Miss Addie Eliza Ritchie, on March 9, 1882, at Scranton. Here they purchased a farm and resided there until they moved to a farm near Rockwell City. In 1916 they retired and moved to Hampton, where Mrs. Fitz died in 1925.

Surviving Mr. Fitz are three daughters and five sons, Miss Myrtle Fitz, at home; Mrs. Velma Heideman, of Rockwell City, Mrs. Hazel Marken, of Waterloo, Charles Fitz and Ray Fitz, of Dows, Earl Fitz, of Fayette, Lloyd Fitz, of Independence, and Ralph Fitz, of Alexander; also sixteen grandchildren and two great grandchildren. Two children preceded him in death.

Grandpa Fitz's death notice.

CHAPTER 5

WAR CLOUDS GATHER

While the assassination of Archduke Franz Ferdinand of Austria during a visit to Sarajevo on June 28, 1914 was the immediate cause of World War I, tensions had been heating up in that part of Europe for some time. In short order Austria-Hungary declared war against Serbia believing, probably correctly, that Serbia had played a role in the assassination, and Russia, being an ally of Serbia, started mobilizing its army. In short order, essentially all of Europe would be involved in the conflict. By the end of the war more than 20 countries had declared war. Germany, Austria-Hungary, Bulgaria and the Ottoman Empire on one side and Great Britain, France, Russia, Italy and Romania on the other, and finally Japan and the United States.

President Wilson had taken office on March 4, 1913 and so was settled into his first term by the time hostilities broke out in Europe. Within a little over a month after the war started however, the President's wife died, leaving him lonely and depressed. While he continued to handle his official duties, he had no stomach for war.

The official position on the war taken by the United States under President Woodrow Wilson was one of neutrality. It was a European war to be dealt with by the Europeans the President believed. Wilson declared, "the United States must be neutral in fact, as well as in name, during these days that are to try men's souls."

Ralph had been itching for some time, we think, to get started farming on his own. He was 23 years old and he was restless.

The time period in Montana of which we are writing was both exciting and sad. In the first half of the 1800s, central Montana was cowboy country with large cow herds being pastured free range and with cowboys blowing off steam in the surrounding towns that were liberally supplied with saloons and other business's catering to lonesome cowboys. In the early days of Montana Territory, the area from the Missouri River north to the Canadian border and from the North Dakota line westward to the foothills of the Rocky Mountains was an Indian reservation (according to an 1883 map it was the Gros Ventre, Piegan Wood, Blackfeet and River Crow Indian Reservation).

This would have included the area that became the town of Whitewater and the land which Ralph would eventually homestead. When this reservation was opened to white settlement the big cow herds expanded to include this area in their free range. Then came the winter of 1886-87 which William McGee described in his book *Montana Memoir* in excerpts from *Forest Grove, Montana: A History Book*, with the following passage:

> Then in the fall of 1886, over supply and low prices, forced many cattlemen to hold over their fat cattle. Winter struck early, heavy wet snow, then a chinook melted snow over 50 square miles [a chinook wind, popularly referred to as an 'ice-eater,' is a warm wind that melts snow.] Blizzards turned water into sheets of ice, more snow, then blizzards, cold winds, and 30 to 40 below zero for weeks. The cattle died, not by hundreds, but by hundreds of thousands. (12) pg 3

That pretty much ended the era of the massive free-range cattle herds and when the severe loss of cattle again occurred during the winter of 1906-07 even the small cattle operators that were operating in the opened Indian reservation could not survive. That left the area to the homesteaders and James J. Hill was ready to take advantage of the situation.

James J. Hill among other ventures, was primarily a railroad builder. He had a dream to build a northern railroad line from St Paul, Minnesota to Seattle, Washington. Originally called the St Paul, Minneapolis and Manitoba Railway, it was later named the Great Northern Railway and in Montana became known as the Montana Hi-Line. Once the railroad was completed, Hill set out to populate the area and establish business for the railroad. This he did, in part, with a widespread advertising campaign in newspapers and public buildings extolling the great opportunities for homesteaders in Montana. The newspapers, in addition to the stories about the war in Europe, were printing stories and advertisements about the opportunities in Montana. Had Ralph been reading these advertisements in the newspaper? Maybe. Probably? No one else in his family had ventured that far west.

From the Fort Laramie Treaty of 1851 onward, the area that would become the state of Montana in 1889 was home to various Indian tribes. As the frontier was pushed further west by settlement, treaties were negotiated and presidential executive orders were issued that established reservations for each of the tribes. Allotments were made to the tribal members and after each tribal member was assigned their allotment of a designated number of acres, the surplus lands were opened for homesteading. This became known as checkerboarding and Ralph would soon become part of it.

The United States had inherited the concept of the yeoman farmer from Great Britain and as early as Thomas Jefferson, had been attempting to pass some form of homestead act, but the southern states, believing that it would threaten their system of slavery, were able to block passage. During the Civil War however, that obstruction disappeared and President Lincoln was able to sign the first Homestead Act of 1862. In 1909 the Enlarged Homestead Act was passed which offered homesteaders 320 acres of land.

While life in the United States was only moderately affected, the fighting was fierce in France. Stories on the trench warfare appeared daily in U.S. newspapers. Unbeknown to all but a few British military of high rank, the British had Room 40. Early in the hostilities, the English had found a copy of the German code books in the aftermath of a battle, and Room 40 was busily decoding German communication to the great advantage of the Allies. The war, at least the land war and at least in part because of Room 40, was not going well for the Germans. They were beginning to look toward the war at sea as a way to defeat the Allies, especially the British.

While both Britain and Germany thought the sea war would be won or lost by battleships, the Germans started looking to the submarine, which had not been used as a weapon of war until now, as an effective weapon. The submarines that the Germans were building had the advantage of stealth since they could travel under water for long distances by battery power only having to surface from time to time to run the diesel engines and recharge the batteries. When submerged, the only evidence that they were in the vicinity was the periscope protruding above the water, which from a distance looked like a tin can floating on the waves, and the narrow wake trail made by the periscope as it moved through the water. The submarine could, hopefully undetected, spot a target ship through the periscope and discharge a torpedo without having to surface. Only if they wanted to use the guns mounted on their deck or recharge their batteries would they have to come to the surface. They were slow moving compared to steam driven ships but they had the advantage of being able to sneak up on their target.

People of the United States supported their President and in large part were in support of the stance of neutrality that the President was advocating. By May 1st, 1915, May Basket Day for the children, the luxury liner *Lusitania* left the dock on the Hudson River in Manhattan bound for Liverpool, England. The President was busy and in between

Lusitania *leaves New York Harbor.*

official duties was courting the Washington widow, Edith Galt.

The *Lusitania* was a luxury ocean liner owned by the Cunard Company of Britain, the largest in the world, with four steam boilers powered by coal.

R.M.S. Lusitania *(Cunard Line).*

With enough men to shovel coal the liner could travel about 30 miles an hour, faster than most other ships and twice as fast as a submarine. She was a magnificent ship, the choice of the rich and famous who wished to visit England. The Lusitania had made the trip across the Atlantic 201 times as of May 1915, but this would be her last. The *Washington Times* had warned of trouble ahead when they wrote,

> The liner Lusitania, with several hundred prominent Americans on board, is steaming toward England despite anonymous warnings to individual passengers and a formally signed warning published in the advertising columns of American newspapers — warnings, which, in view of late developments in the sea war zone, it is beginning to be feared, may prove far from empty.
>
> ...Hundreds of Americans are holding their breaths lest relatives on board such vessels go down... ... But what is worrying everybody is the accumulation of evidence that Germany is either looking for trouble with the United States or that her authorities are reckless of the possibilities of incurring trouble.

But it was full steam ahead for the Lusitania. After all, there were the unwritten rules of warfare that you do not attack civilian ships and passenger ships were just off limits. Even Winston Churchill thought the idea that submarines would attack merchant ships was absurd. In times of war, however, accepted rules are subject to being ignored. In fact, German policy vacillated between authorizing unrestricted attack on merchant ships and not authorizing it, depending on who was in command, especially when it came to submarines, called U-boats.

As the Lusitania got within a day's sailing of the dock in Liverpool, England, Captain Turner slowed the pace in order to arrive at the dock at the assigned time the next morning. At about the same time the ruthless German submarine Commander Schwieger was cruising in the area with submarine U-20 looking for merchant ships he could sink to add to his total of tonnage sunk. Room 40 had been tracking U-20 and knew the sub was in the same area as the Lusitania. With all these warnings, no escort was pro-

Lusitania *graves – two graves, 100 bodies per grave.*

vided. As the Lusitania made a turn, just off the coast of Ireland, she came into near perfect alignment and Schweiger fired his most powerful torpedo.

Twelve hundred Lusitania passengers breathed their last breath as a result with 128 of them being Americans, and the Lusitania sat on the ocean floor. While the citizens of the United States were becoming increasingly outraged, and former president Teddy Roosevelt led a faction wanting war, most Americans were not ready for war. President Wilson issued a strong protest against the sinking of the Lusitania, but still favored the position of neutrality.

At least for now the United States was staying out of the war, it was business as usual and Ralph left the family farm near Rockwell City, Iowa by train on March 23, 1917 for Lewiston, Montana. His intent, we believe, was to get a start in farming, but his trip was a bit delayed by a train wreck in North Dakota. He arrived in Harlowton Saturday evening, went

North Dakota train wreck.

Mike Nettik's farm.

to church Sunday morning and found a job working for Mike Nettik near Hilger, Montana on Monday. Ralph's first impression of farming in Montana was while working for Mike Nettik and while his stay there was rather brief it seemed like a relatively successful

Judith Mountain view from Nettick farm.

farming operation. The scenery was beautiful with the Judith mountains in the distance where mining for gold and other minerals had attracted attention only a few years previous. It was different than the Iowa farm he had left but still exciting to be starting a new adventure. Ralph, and most likely most Americans, were unaware of what had been detected in Room 40 the previous January.

While President Wilson was of the opinion that the United States should stay out of the war, Germany would make that option impossible. Quoting a graphic display at the National WW I Museum and Memorial in Kansas City,

> On January 16, 1917, Germany's foreign minister, Arthur Zimmerman, cabled a message to Count von Bernstorff, Germany's ambassador in Washington, D.C. Bernstorff then relayed the telegram to Mexico. The British intercepted and

decoded the message, discovering the German proposal for an alliance with Mexico against the United States.

Essentially Germany was saying to Mexico, if you will help us defeat the United States, we will make sure you get Texas, New Mexico and Arizona back.

Less than two weeks after Ralph had arrived in Montana and started working for Mike Nettik, the United States declared war against Germany on April 6, 1917.

Bunkhouse

At first Ralph slept in the house with the Nettiks but on April 4th he moved his bed roll to the bunkhouse, but continued to have meals with the Nettik family. Throughout April the weather brought heavy wet snow storms followed by warm days and melting snow. This took a toll on young calves and even some cows who got mired in the snow and died. It was Ralph's job to skin the cows and calves that died so as to at least salvage the cow hides. When not skinning cattle Ralph was hauling straw for the cat-

Skinning calves

tle, helping mired cattle out of the snow, fixing and greasing harness, building fence, hauling manure and picking rock. By the first part of May the field work was starting and Ralph was busy discing, harrowing, plowing and drilling seed.

As the newspapers spread the word concerning the happenings in the rest of the world, a ground swell was developing across the country in support of the war effort. In his diary Ralph writes that there was a patriotic celebration in Hilger, Montana on May 8 but he makes no mention of attending, only that he disced in the forenoon and harrowed alfalfa in the afternoon.

Promoting the War

On May 18, 1917 President Wilson signed the Selective Service Act which required all men between the ages of 21 and 30 to register for the draft. On June 5 Ralph went to Hilger and registered for the draft as he was

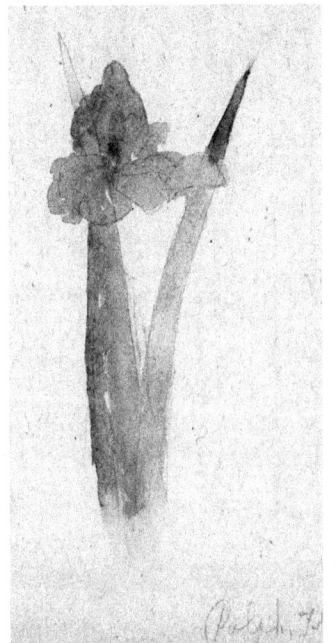

Ralph's painting of a lily and another of a church.

Draft registration.

required to do. June 10th, a Sunday, he wrote in his diary, it rained all day today. I made pictures.

He, no doubt, was also deep in thought weighing his options. This war was getting in the way of his plan to get a start in farming and yet he read in the newspapers about the opportunities for homesteading and even that there were special opportunities in homesteading if one were to serve in the military. But he also read about the war in France. In spite of President Wilson's attempt to censor the press in reporting of the war news, he had heard about the conditions in the trenches in France and about that wicked barbed wire and the casualties, both the death toll and the wounded. What should he do?

His baby sister Hazel, who was eleven at the time, wrote him saying, "I hope you don't have to go to war because if you do I'm afraid you will get killed." But then just in case, she wrote, "What are you going to do with your camera when you go to war? I can tell you send it to me. ha! ha!"

It is obvious, we think, that Ralph was not aware of what was actually happening to homesteaders in Montana. Like those before him he would have had primarily only the newspaper for guidance, a newspaper that was focusing on the war in Europe and the revenue generated by the advertising of homestead opportunities in Montana. Monday morning quarterbacking is easy but William McGee, quoting from *Forest Grove, Montana: A History Book,* in his book *Montana Memoir* years afterward, described it this way.

"Between 1900 and 1918, 80 to 90 thousand flooded into eastern Montana. In good times, the homesteaders plowed in the spring, prayed for rain, harvested and hauled to the granary in the fall, and then went to the banker to borrow more money, for more land and seed. Loans exceeded deposits by 50 percent. Most of the loans on next year's crops. The price of wheat was good. The towns and competition were growing. (12) pg 3

...

The plains did what the plains had done for millions of years. They went through their cycle of drought and moisture. It began in the spring of 1917, when rain registered .33 of an inch in May and .45 in June, the wet months. The drought spread and by 1919, covered all of central and eastern Montana. Hot sustained winds, mild winters, rolling range fires, and grasshoppers. Endless clouds of black, ugly, clacking, whining insects eating every blade of dry grass, trees, and shingles. The marginal homesteaders fought to exhaustion and gave up. The exodus began by every possible means, taking with them what little they had left. Leaving behind dying towns, houses that weathered and fell down, and broken hearts. Banks failed with millions of dollars in mortgages, indebtedness, and millions of acres of eroded land. By the early 1920s, 60 to 70 thousand had left Montana." (12) pg 4

Floyd Hardin in his book *Campfires and Cowchips* tells a story descriptive of homesteader woes. While riding horseback they observed what they described as follows:

> About midway of the sag stood a homestead shack. As we drew near we saw it was deserted. The single window, glass broken, stared vacantly across the dog town [prairie dogs], as if morosely contemplating the past. The tar paper roof was punched full of holes, the result of some hail storms. The one by twelve plank floor, broken in many places by range stock seeking protection from flies and heat, was a nesting place for pack rats and littered with sage brush branches and prickly pear which these industrious little pests had dragged in. Openings to the skunk dens under the broken floor gaped starkly from under the flimsy foundation. The single door, half open, sagged resignedly from one hinge, as if it had at last become weary of the battle against the elements and range stock and given up the fight. The corners of the shack, smooth and oily from much rubbing by cattle and horses, glistened in the sunlight, while here and there a splintered board flapped idly in the breezes.
>
> A closer inspection revealed a weathered piece of cardboard tacked to the sagging door. A faded inscription, printed with blunt lead pencil vividly pictured the disappointment and disgust of the cabin's former occupant with the words, "Left to the skunks and coyotes. Farewell, Montana, Farewell!" (11) pg 80

A poetic description of homesteading in Montana and a far cry from the Great Northern Railway newspaper advertisements: Free land to any homesteader not afraid to work.

But Ralph, we think, did not know any of this, and not until he returned from his enlistment in the Navy and the end of the war did his future in Montana become clear to him.

By June 11th Ralph had sorted things out and wrote to his father writing, "I don't think I will start farming this spring. I see in the paper there is an Indian Reservation open to homestead in the northern part of the state. It is just north of the Missouri River. There is 40,000 acres of it. I believe I will go up there and see what kind of land it is, and if it is worth taking get me a homestead and then join the Navy. I wouldn't have to do anything with it while I was in the Navy. And if I don't join the Navy I probably will be drafted and have to go to France and I don't think much of that."

In his diary for June 13 and 14 Ralph wrote, "Weather fine. I hauled manure." On June 15th he quit working for Mike Nettik and went to Lewistown, perhaps just to make his plan without the pressures of his job, but the next day he went back to Hilger to settle up. The next day was Sunday and Ralph was back in Lewistown. Monday morning, he went to inquire about joining the Navy and then started for Glasgow to look at homestead land arriving there the morning of the 19th. He found a locator, Mr. Hawley, and they started out to look at homestead land, staying overnight in Saco the first night out.

Ralph and Mr. Hawley left Saco mid-morning the next day, looked at two or three pieces of land some thirty miles north of Saco and then headed cross country for Whitewater. Within a mile after starting their trip to Whitewater Ralph writes in his diary, "… we came to a rock in the trail. Mr. Hawley turned out to miss the rock and when he did so the gears locked in low and he could not stop till we jumped over a ten-foot bank. The car stood straight up and down in the mud and water. We got a man and team to help us, and it took eight hours to get out. Nothing was broke on the car." The next day they worked on the car until about four in the afternoon but could not get it started. Frustrated, Ralph found a man and team who pulled them three miles, probably trying to get the car to start, but without any luck, and so they gave up for that day. The next day the same man pulled them into Saco behind his wagon. By nine o'clock that night they had gotten the car started and headed for Mr. Hawley's homestead where they spent the night.

Ralph's Homestead

Three days they had spent trying to travel the thirty miles between Saco and Whitewater and finally they were there. Mr. Hawley showed Ralph a 320-acre homestead about 2-1/2 miles from the Whitewater Post Office. It was a relinquishment which meant that another homesteader had homesteaded it, but then for unknown reasons had relinquished it back to the government. Ralph was ready. He purchased the homestead for $500. It must not have taken him very long to make up his mind because he says in his diary that they were back in Saco before noon.

Whitewater, Montana was essentially just the Whitewater post office and general store and got its name from the cloudy appearance of nearby Whitewater Creek. In 1927, N.J. Brandt, the owner of the store and post office moved it, and with it the town, seven miles west to be on the Great Northern Railway line.

Town of Whitewater.

That afternoon in Saco, Ralph wrote to his father, partly to ask his father to send the money he needed to pay for the homestead he had just purchased, but also a "what if" letter. A dreaming letter of a young ambitious farmer. If only this war weren't in the way. He wrote,

> … I was thinking about joining the Navy but if I could start farming I don't think I would have to go to war. I don't hardly know what it would cost to start farming a little. If I had 3 horses and a plow and disc and harrow and wagon I could put in a quite a bit of grain. I could borrow or hire a drill and it wouldn't cost me very much. I don't know whether I could borrow any money out here where nobody knows me or not. I thought if I could get some horses I would go out and start plowing about the first of August. Then I could put some in fall wheat and the rest I'd get plowed in spring grain. … I don't know what I could get horses for. Good horses are pretty high out here. I could plow a lot with two horses and a walking plow. … I think I got a cheap piece of land. …

The next day which was Sunday he just stayed around Saco and Monday morning started walking to his new homestead. "I walked about 13 miles" he wrote in his diary, "then caught a ride in a Ford the rest of the way." For 5 or 6 days he spent living at and getting acquainted with his homestead, no doubt weighing his options for the future, and waiting for a letter from his father with the money he needed to complete the homestead filing. On the last day of June, he helped one of his new neighbors build fence and the next day walked to Saco to see if his letter had arrived from his father. He stayed in Saco the next day and apparently received the letter and money from his father. On July 3rd he went to Glasgow, paid for and filed on his 320-acre homestead and also filed on an additional 80 acres, and then returned to Saco. July 4th was a holiday but no celebration in Saco.

Ralph was tormented. He didn't want to go to war to fight Europe's war. He wanted to start farming. July 5th, he walked to his homestead and at 8 o'clock that evening started walking back to Saco. He walked all night and arrived in Saco at 4:30 the next morning. He took the 10 o'clock train to Lewiston and on to Great Falls where he took the examination for the Navy

288

HOMESTEAD ENTRY AND PATENT RECORD No. 7

Doc. No. 98684

G

Glasgow 050004 and 050026   4-1003.

The United States of America,

To All to Whom These Presents Shall Come, Greeting:

WHEREAS, a Certificate of the Register of the Land Office at ....Glasgow..., Montana, has been deposited in the General Land Office, whereby it appears that, pursuant to the Act of Congress of May 20, 1862, "To Secure Homesteads to Actual Settlers on the Public Domain," and the acts supplemental thereto, the claim of................................

........Ralph G. Fitz........

has been established and duly consummated, in conformity to law, for the ....

........north half of Section thirty-three in Township thirty-five north........

........of Range thirty-two east of the Montana Meridian, Montana, contain-........

........ing three hundred twenty acres,........

according to the Official Plat of the survey of the said land, returned to the General Land Office by the Surveyor-General:

NOW KNOW YE, That there is, therefore, granted by the United States unto the said claimant......................

the tract of Land above described; TO HAVE AND TO HOLD the said tract of land, with the appurtenances thereof, unto the said claimant and to the heirs and assigns of the said claimant forever; subject to any vested and accrued water rights for mining, agricultural, manufacturing, or other purposes, and rights to ditches and reservoirs used in connection with such water rights, as may be recognized and acknowledged by the local customs, laws, and decisions of courts; and there is reserved from the lands hereby granted a right of way thereon for ditches or canals constructed by the authority of the United States.

IN TESTIMONY WHEREOF, I, ...Warren G. Harding........, President of the United States of America, have caused these Letters to be made Patent, and the seal of the General Land Office to be hereunto affixed.

Given under my hand, in the District of Columbia, the........nineteenth........day of

........May........in the year of our Lord one thousand

SEAL

nine hundred and........Twenty-one........and of the Independence of the

United States the one hundred and........forty-fifth.

By the President:........Warren G. Harding........

By ....M. P. LeRoy........, Secretary.

Recorded: Patent Number........807054........

6-6020

........L. Q. C Lamar.........
Recorder of the General Land Office.

Filed for record this....14th....day of........April........A. D. 1930, at....9....o'clock....A.M.

........J. F. Roadhouse........ Clerk and Recorder.

By........................, Deputy.

Purchased 320-acre homestead.

289

PHILLIPS COUNTY, MONTANA

Doc. No. 98685

G

Glasgow 050004 and 050026   4-1007.

The United States of America,

To All to Whom These Presents Shall Come, Greeting:

WHEREAS, a Certificate of the Register of the Land Office at ____Glasgow_____, Montana, has been deposited in the General Land Office, whereby it appears that, pursuant to the Act of Congress of May 20, 1862, "To Secure Homesteads to Actual Settlers on the Public Domain," and the acts supplemental thereto, the claim of _____

_____Ralph G. Fitz_____

has been established and duly consummated, in conformity to law, for the _____

_____northwest quarter of the southeast quarter and the northeast quarter_____

_____of the southwest quarter of Section thirty-three in Township thirty-_____

_____five north of Range thirty-two east of the Montana Meridian, Montana,_____

_____containing eighty acres,_____

according to the Official Plat of the survey of the said land, returned to the General Land Office by the Surveyor-General:

NOW KNOW YE, That there is, therefore, granted by the United States unto the said claimant_____

the tract of Land above described; TO HAVE AND TO HOLD the said tract of land, with the appurtenances thereof, unto the said claimant and to the heirs and assigns of the said claimant forever; subject to any vested and accrued water rights for mining, agricultural, manufacturing, or other purposes, and rights to ditches and reservoirs used in connection with such water rights, as may be recognized and acknowledged by the local customs, laws, and decisions of courts; and there is reserved from the lands hereby granted a right of way thereon for ditches or canals constructed by the authority of the United States. Excepting and reserving, however, to the United States all the coal and other minerals in the lands so entered and patented, together with the right to prospect for, mine, and remove the same pursuant to the provisions and limitations of the Act of December 29, 1916 (39 Stat. 862).

IN TESTIMONY WHEREOF, I,_____Warren G. Harding_____, President of the United States of America, have caused these Letters to be made Patent, and the seal of the General Land Office to be hereunto affixed.

SEAL

Given under my hand, in the District of Columbia, the_____nineteenth_____day of _____May_____in the year of our Lord one thousand nine hundred and_____twenty-one_____and of the Independence of the United States the one hundred and_____forty-fifth.

By the President:_____Warren G. Harding_____

By_____M. P. LeRoy_____, Secretary.

Recorded: Patent Number_____807055_____

6-2176

L. Q. C Lamar
Recorder of the General Land Office.

Filed for record this____14th____day of____April____A. D. 19_30_, at____9____o'clock A. M.

Indexed ES
Compared
ERM
Paged ES

J. T. Roadhouse Clerk and Recorder.

By_____Deputy.

Filed on 80-acre homestead.

and passed. Then back to Lewistown to get the suitcases he had left there and turned around on Sunday the 8th for a return to Great Falls. In Great Falls Ralph made arrangements with the Navy officer to get to Salt Lake City for the final examination for the Navy. Then back to Saco for final preparations for being away for a while.

On July 11,1917, Ralph and two other fellows left for Salt Lake City. There he took another examination, which he passed, and was sworn into the United States Navy. He was now a sailor in the United States Navy. Farming would have to wait.

CHAPTER 6

SERVING HIS COUNTRY

Over there, over there,
Send the word, send the word over there
That the Yanks are coming,
The Yanks are coming,
The drums rum-tumming Ev'rywhere.
So prepare, say a pray'r,
Send the word, send the word to beware.
We'll be over, we're coming over,
And we won't come back till it's over
Over there.

Song written by George M. Cohan, 1917

On July 12, 1917, Ralph left Great Falls, Montana on the train bound for Salt Lake City and arrived there at nine o'clock in the morning. He did a little sight-seeing in Salt Lake City, toured the Mormon Tabernacle and the surrounding grounds, visited the pavilion on Great Salt Lake and then headed for the recruiting office where he took the examination for the Navy and passed. He enlisted and was sworn into the U.S. Navy on July 12, 1917 in Salt Lake City, Utah.

The next day, back on the train headed for Bremerton Navy Yard in Washington, by way of Pocotello, Idaho and Portland, Oregon, arriving in Portland at 7:00 in the evening on the 14th and then on to Seattle, Washington. Seattle is on the east side of Pudget Sound and the Bremerton Navy Yard was on the west side so back on the train traveling around the southern tip of

Entrance to Puget Sound Navy Yard.

the sound and arriving at the Bremerton Navy Yard on July 15. Ralph drew his blankets and mattress and got his tent assignment in the detention camp along with tentmates Beaver and Thayer where they expected to stay for the next twenty-one days. Three days later Ralph stood his first sentry duty.

Bremerton Navy Yard was first established in 1891 as a repair facility but in 1916 became a major ship building facility. In World War I they built submarine chasers, submarines, mine sweepers, tugboats, ammunition ships and many small boats.

On August 10 Ralph was moved from Detention Camp to Camp #2. The next day he was given liberty but only long enough to attend church. While Ralph makes no mention of what was occurring in Detention Camp and Camp #2, we think that they were most likely kept busy with training and indoctrination into Navy life. That may have been when he was issued the Bluejackets Manual, United States Navy, 1917. Three and a half months would go by before on October 27, 1917 Ralph drew his hammock and packed up ready to leave Bremerton Navy Yard.

The Bluejacket Manual, 1917.

On October 28 they rose early, had chow, and marched to the dock with the band in the lead playing as they marched. They loaded on a tugboat and shoved off across the sound for Seattle. They arrived in Seattle at about 9:30 in the morning and waited for the train, the first of two trains, to arrive. Ralph, being always mindful of meals on time, made note of the fact that it was three in the afternoon when the first train pulled in, and they had had nothing to eat since an early breakfast. But at least he was lucky enough to be on the first train of 200 sailors, and he was happy about that. The second train was about a half hour behind. Ralph makes note of the fact that the train had 200 sailors on board and only had two cooks, so it was nine o'clock that night before he got anything to eat. Finally, on a full stomach he crawled in bed at eleven.

The next morning found the train in Montana. For a full week they traveled aboard the train, east from Washington through Montana, North Dakota and South Dakota and on to Minneapolis, Minnesota where they changed and headed south through Iowa to Dubuque and then east again to Chicago, Illinois where they stopped. Forty-five minutes later they proceeded eastward through Indiana, Ohio and West Virginia, arriving in Norfolk, Virginia at 6 am on November 2.

After marching about half a mile, carrying their bags and hammock, they arrived at the barracks. That would be their first night sleeping in their hammocks and Ralph did not fall out of his hammock, maybe because every little while he would be awakened by a thud as one of his comrades rolled over in the hammock and landed on the floor. Ralph says in his diary, "We had a lot of fun."

November 3rd was spent cleaning and rolling the streets and Ralph said the "Grub [was] punk, not enough to eat." The next day after breakfast at six, they took the boat to the USS *Illinois*. Again, Ralph comments, "Had nothing to eat all day." After supper that evening Ralph was assigned to his billet, which is simply the quarters where he was assigned to sleep. Ralph was apparently not impressed. He commented, "This is sure a mad house, not even room to sit down."

Already the next day Ralph had his first duty on board ship working in the dynamo [electrical generator] room. The following week was spent in lessons in firing the weapons onboard the ship.

U.S.S. Illinois

The USS Illinois had been launched in October 1898 and following her service in various fleets she was used as a training ship from November 1912. Following a 58-year career she was sold for scrap in 1956. Following the week of lessons in firing the guns onboard, the USS Illinois spent the next week out at sea. Ralph commented, "...sure was seasick. Nearly every man in the training draft was seasick ..." Ralph was assigned to compartment cleaning that week. The next week he was assigned to the fire room which was where the boilers were located that propelled the vessel. And then as November was coming to a close his luck changed and he was assigned for the week to mess cooking for the Chief Petty Officers. "Am getting fine eats." he wrote in his diary. "Lots of pie and cake and puddings and all the sugar and milk I can eat."

By mid-1917 the first American soldiers were entering combat in France and on the seas the convoy system was becoming common practice in guarding ships sailing through the war zones.

On the morning of December 16th Ralph and his comrades packed at 10 am ready to leave for Boston, Massachusetts that evening at 6 pm. They were scheduled to leave on a small passenger ship bound for Norfolk, Virginia. They sailed most of the night arriving at the Norfolk Navy Yards at 2 am having gotten no sleep because there were no hooks on which to swing their hammocks. By nine o'clock they were taken to the mess hall and put to work peeling spuds [potatoes] only to get the word at 10:30 to get into dress blues and be ready to fall in with bags and hammocks. At 3 pm they boarded

*Charleston Navy Yard located in the Charleston district of Boston,
home of all naval officers and where all warships coming to this port to find a
place to dock.*

the ship for Boston and sailed at 6 pm. The next day, December 18th, Ralph writes, "We are out at sea now. The sea is not very rough but this small ship is rocking badly. Most of the boys are seasick. Some are not able to eat at all. I am getting pretty dizzy." The next day they sailed into Boston in the afternoon and marched to Commonwealth pier with their bags and hammocks. For the next week Ralph was part of a working party to the Charleston Navy yards and Ralph worked most of that time on the USS Shawmut. In 1917 the ship renamed the USS Shawmut had been purchased and converted to a minelayer to be used to lay mines in World War I.

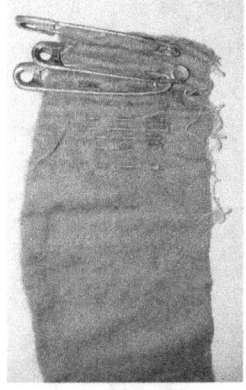

Sewing kit

Matches

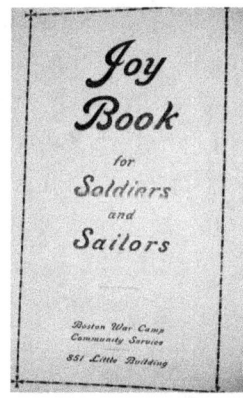

Keeping morale up at Boston War Camp

Buddies during the winter of 1917 while stationed in Boston.

Christmas Day 1917 was made special for the sailors. They had a Christmas tree and Santa Claus gave every sailor a Red Cross bag of candy and other useful articles which probably included a sewing kit and may have included a box of matches and a book to boost their morale. In the afternoon there was dancing with visitors allowed so lots of local girls had their chance to dance with a sailor and in the evening a good show. Ralph writes in his diary, "They sure did give us some turkey dinner to (sic), and ice cream to finish up on."

From January 10th to the 28th Ralph had been mess cooking and on the 28th got liberty until March 2nd.

Mid-morning on March 2nd, they left Boston for New York City. Arriving there at 6 pm, they ate at a restaurant and were taken to the receiving ship at Brooklyn Navy Yard by motor truck, traveling over the Manhattan Bridge which Ralph described as "the longest bridge in the world." The next day they were transferred from the receiving ship to the USS Leviathon. This ship which had been renamed the USS *Leviathon* was a German luxury liner named the SS *Vaterland*, the pride of the German passenger fleet, before she was renamed. She was anchored in New York when the United States entered the war and the United States took her prisoner and put her to use as a troop transport. She was the biggest ship in the world and Ralph wrote in his diary, "This is sure some ship. It has 14 decks and 46 boilers." History shows that a few weeks after Ralph had been on the ship, she was stricken by the Spanish Flu and many soldiers were buried at sea. On March 4th the

Receiving ship, New York, March 1917. Ralph is on the left.

Receiving ship, New York.

Leviathon sailed out of New York bound for Liverpool, England and Ralph went on watch in the fireroom from 8 until noon that day and took shifts in the fireroom for the next four days.

Following his watch in the fireroom, he had no assignment for a couple of days. "Nothing to do," he wrote. But the torpedo boats that were convoying the USS *Leviathon* had something to do. They had sighted a submarine about 6 am on the 10th of March. They had dropped two depth bombs and the submarine was reported sunk. The next day Ralph was on watch in the fireroom again and the following morning they sailed into Liverpool, England. Training was over. They had survived the trip to Europe and were now at war.

They left the USS *Leviathon* at about 2 in the afternoon, got a free supper at the YMCA in Liverpool and had liberty until 9 pm when they boarded the

USS Leviathon, *the ship Ralph went across on.*

train at Line Street Station bound for the major Irish seaport at Holyhead, Wales. There they boarded an old cattle boat and sailed across the Irish Sea to Dublin, Ireland. They arrived in Dublin the next morning and Ralph wrote, "Almost froze coming over." After a free lunch at the soldiers and sailors free lunch counter it was back on a train bound for Queenstown, Ireland. Arriving in Queensland at

Depth bomb explosion.

5:30 in the afternoon, they were served supper on the USS *Melville* and then transferred to the barracks at Passage, Ireland where Ralph would remain until April 8th. Word spread rapidly through the barracks that the cattle boat that had brought them across the Irish Sea had been torpedoed and sunk on her way back to Holyhead. They had survived another near miss.

Ralph spent the next few days mostly in a work party on the USS *Camden*. The USS *Camden* was a cargo ship that had been seized from Germany and was being used to transport supplies and Ralph was part of the detail that was unloading her. He spent Easter Sunday working on the USS *Camden* and wrote in his diary, "No Easter eggs for the sailors today." By April 3rd the USS *Camden* was ready to sail for the United States, but first they held funeral services for 16 victims of the USS *Manley* disaster, whose bodies they would transfer back to the United States.

Taking the victims of the U.S.S. Manley *ashore at Queenstown, Ireland, 1918.*

U.S.S. Manley *was blown up by her own depth charges after being rammed by a British ship.*

On March 19th off the coast of Ireland the USS Manley was acting as part of a convoy. The disaster had happened when the stern of the USS Manley collided with the merchant cruiser HMS *Motagua*, causing the USS *Manley's* depth charges to explode. The explosion sent shrapnel into the gasoline and alcohol tanks which burst into flame, the explosion and fire killing 34 men on the USS Manley and 28 on the HMS *Motagua*.

On April 8, 1918 at about 1 pm Ralph left the Barracks at Passage, Ireland for the USS *Allen* which was anchored in the harbor off Queenstown, Ireland, the ship on which he would remain for the remainder of the war. Ralph had a diddy box in which to keep his personal possessions and we suspect that each sailor had been issued one. The next day he would first experience the convoy mission of his ship, as they escorted a ship through the war zone and returned to Queenstown 3 days later.

The USS *Allen* was a Sampson class destroyer who served out of Queenstown, Ireland during the first World War, a

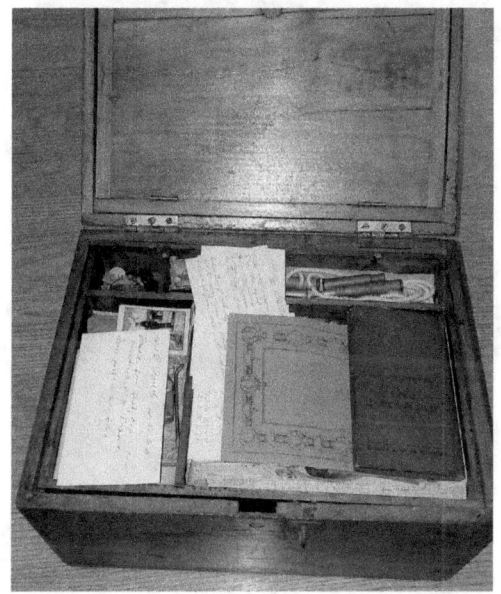

Diddy box.

destroyer of some renown. She was launched December 5, 1916 and commissioned January 24, 1917, after which she would leave New York on June 14, 1917 for Queenstown, Ireland where she would be based. She was under the command of Lt. Commander Samuel W. Bryant and her primary purpose was to be used on anti-submarine patrols and convoy escort duties. She was equipped with four 4in/50 guns, two one pounder anti-aircraft guns and twelve 21-inch torpedoes as well as depth charges.

The USS *Allen* carried out ten attacks on suspected U-boats (submarines). The sailors who served on the USS *Allen* from June 28, 1917 through November 11, 1918 qualified for the First World War Victory Medal. We are not aware that Ralph ever received that medal but perhaps he would have if he had stayed in the Navy after the war was over. She survived to be the only one of the 1,000 tonner destroyers to see service in World War II and was present at Pearl Harbor

USS Allen *in Queenstown, Ireland.*

Row boats from USS Allen.

Motorboat No. 66 from USS Allen.

during the Japanese attack on December 7, 1941. On September 26, 1946 she ended her service and was sold for scrap.

On the USS *Allen*

On the USS *Allen*

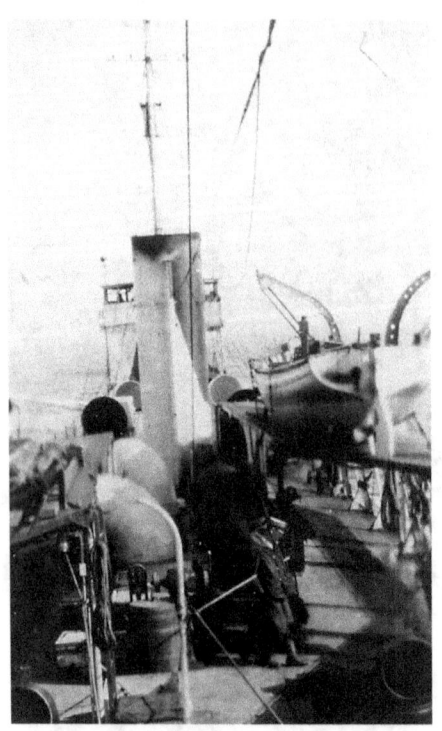

USS Allen, *taken from forward mast lookout*

Rapid fire gun, USS Allen.

USS Allen *cooks.*

Shipmates on USS *Allen*, 1918–1919

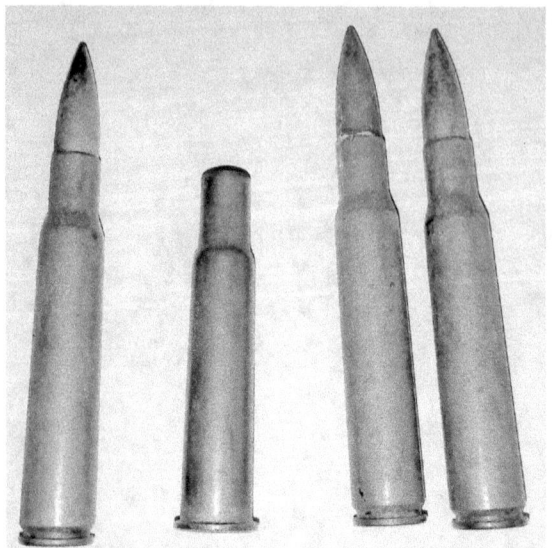 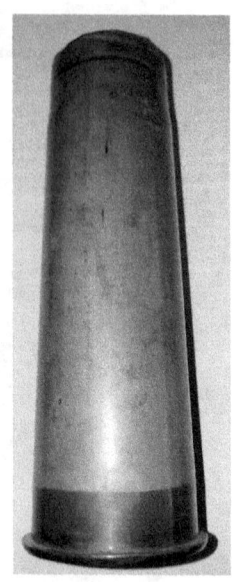

Unfired ammunition from USS Allen.   *Shell casing from a bigger gun.*

On April 23rd an enemy submarine was spotted at 7:45 in the morning. Ralph wrote in his diary, "We immediately headed straight for it firing two shots at her from the bow guns. When she submerged we then dropped three depth bombs on the spot where she submerged and then circled around and dropped sixteen more."

Depth charges.

Depth charges on USS Cushing.

Depth bomb from USS Cushing.

By the time of the First World War most destroyers had switched from coal fired boilers to oil fired boilers to propel the ship. Consequently, the USS *Allen* would need to go into port to take on oil from time to time, which they did on April 24th at Berehaven, Ireland, in preparation for escorting a convoy of some thirty transport ships into Brest, France.

It was a harrowing trip sailing in dense fog for two days which resulted in one ship running into a rock and sinking and another being rammed and sinking. They arrived in Brest on the 28th with their ships that they were convoying, engaged in target practice the next morning and sailed into Queenstown harbor at about ten o'clock.

Depth bomb explosion.

May 6th, they left Queenstown again to convoy three troop ships into Brest, France. They were given liberty in Brest from 2:00 until 9:30 pm. Ralph must have had a good time in Brest. He wrote in his diary, "Brest is quite a town." Ralph got his picture taken in Brest and bought some souvenir hankies for the ladies back home.

By May 7 they were still in Brest and Ralph went ashore at 4 o'clock and got back to the ship at 9 pm. The next day they left Brest to convoy two submarines to Kingston. They got to Kingston at 8 o'clock on the morning of the 10th

Brest, France photo.

Souvenir hankies from France.

and turned right around and headed for Queenstown. Ralph writes, "When we got about to the net we got a wireless that there was a submarine about 12 miles out running on the surface. We immediately turned around and went out full speed. We sighted the submarine and dropped several bombs when she submerged." The nets that Ralph referred to were nets that were placed in the water usually at the entrance to a harbor so that when a submarine got entangled in them a signal would be received that a submarine was in the area. They finally got to Queenstown at 8 am on the 11th, took on oil and anchored on buoy 13. At 5 pm Ralph went ashore and when he returned he was surprised that his ship had been moved alongside the USS *Melville* where it would stay for some general overhauling for the next 5 days.

Upon leaving Queenstown they were patrolling the Irish Sea when on the 19th two British destroyers sighted a sub and started dropping bombs. Ralph writes, "We immediately headed to where they were and dropped several bombs, after which great quantities of oil came to the surface."

On the 21st of May Ralph wrote, "I stood my first auxiliary watch today in the fireroom, which makes me feel very important."

May 23rd they escorted a convoy of 11 ships from Liverpool bound for the United States and then returned to Berehaven to take on oil. Four days later they sailed out and picked up the USS *Leviathon* and escorted her into

Destroyers going out to convoy troop ships through the war zone.

Brest, France, leaving her at the mouth of the harbor and proceeding to Queenstown.

On June 4th they left Queenstown to escort a transport ship out through the war zone and then picked up ten troop ships and brought them back, some going to Brest, France and five with the help of four other destroyers being escorted to Bordeaux, France, after which we returned to Queenstown on June 9th. Two days later they left Queenstown to convoy a fleet of transport ships into Brest, France arriving there on the 15th. The next day they sailed out of Brest to meet a convoy headed for Liverpool, England. On the 20th they left Liverpool escorting a convoy headed for the United States and once through the danger zone they returned to Queenstown. On June 23rd at 1:00 Ralph and others were given liberty to go ashore where they dined on sea pie and fruit salad which Ralph said they enjoyed very much. Sea pie is a pie made with meat or fish comparable to a pot pie.

On the 26th Ralph was mess cooking and the USS *Allen* was convoying a transport out of Queenstown. On the 29th they left the transport out and joined a convoy of 34 transports headed for Brest, France. Upon arriving in Brest, they headed for Queenstown.

July 4th, 1918 was a celebratory day. There was a big doings at the Ringaskiddy grounds in Ireland with baseball games, races, tug-of-war, etc. That may have been where Ralph bought a souvenir hanky. Ralph recorded in his diary, "We had

Souvenir Ireland hanky.

<u>some</u> dinner today, turkey, mashed potatoes, fruit salad, peas, crackers. butter, mince pie. I went ashore to the minstrel show at the Mens Club tonight." The next day they left Queenstown to meet the troop ship Uritania and take her to Liverpool and return to Queenstown where they arrived on the 8th of July.

On July 14th they left Queenstown for Liverpool where they took on oil and then anchored waiting for orders. Ralph wrote, "We had a thunder storm today. First time I have heard thunder since I left Montana over a year ago."

On the 17th they left Liverpool and escorted the *Acritania* out through the war zone. They left her the next day and picked up a fleet of transports on the 19th. On the 20th they left with an oil tanker for Queenstown. At 10:00 that evening they sighted a submarine. They fired one shot and the tanker fired five when she signaled that she was an American sub. Ralph wrote, "She was not hit but the shot fell all around her."

On July 22nd they left Queenstown for Berehaven and got there at 6:00 the next morning. In the afternoon they had torpedo practice. They put wooden heads on them so they would not explode. After they shot one they would follow it up with a motor boat and bring it back. The next day they were still in Berehaven with more torpedo practice and other practice onboard the ship. On July 2nd Ralph wrote, "We have been having gun drill, target practice, abandon ship, etc. the last few days. We left Berehaven today. Are going out to pick up a convoy." They picked up the convoy the next day and the following day sighted a sub at 4:00 in the afternoon. They dropped a buoy where she submerged and circled around and dropped 13 depth bombs. They saw no more signs of her. They arrived safely with the convoy on July 31 and stayed overnight to take on oil.

Back in Queenstown for a few days but on the 8th they left for Berehaven to operate with American submarines for four days. The English sloop, Flying Fox, was with them with stores and depth charges. They were apparently trying to perfect better methods of detecting submarines. On the 12th of August they left Berehaven along with the Kimberley and the Flying Fox. They were trying out some new listening devices but the Flying Fox had also gotten an observation balloon at Berehaven. As they were trying out the balloon on the afternoon of the 14th it fell into the water. The two officers that were in it jumped but both were killed. The Flying Fox got the balloon aboard and went back to Berehaven. The Allen stayed out on patrol on the Irish Sea and English Channel but returned to Queenstown on the 17th to take on oil.

On the 18th they left Queenstown to patrol along the west coast of Ireland and the evening of the 19th they anchored in Blacksox Bay for the night. At noon on the 21st they left for Berehaven and got there on the morning of the 22nd. That afternoon they left Berehaven and Ralph wrote, "Are going to bring in to [two] Battle Wagons." On the 24th they met the USS *Oklahoma* and the USS *Nevada*, two battleships, and escorted them back to Berehaven. They had target practice the next morning and then to Queenstown about 10:30. At 2:00 that afternoon they left Queenstown to follow up on a submarine report some 60 miles outside the nets, and then back to Queenstown the next morning.

On September 9th they left Queenstown for Berehaven to try out a new submarine detector. The next couple of days they worked with the K tube submarine detector but about 1:00 in the afternoon they got a wireless message that a hospital ship, the SS *Galway Castle*, had been torpedoed about 120 miles away. They immediately got under way at full speed and sighted the ship at 7:00 that evening. They approached her at 7:45 and were advised that a British destroyer had picked up all the survivors ahead of them. But they were told that two lifeboats were missing so they started out to look for them. After looking all day and finding nothing they arrived back at the ship and found that four tugboats were starting to tow her in.

That night they lost the hospital ship in the fog and didn't find her again until about noon the next day. One of the tugs had broken loose and did not find the ship again until about 3:00 in the afternoon. The Allen sent a boat over to the torpedoed ship to put a tow line on. They brought back a lot of souvenirs such as blankets, suitcases and such because she was sinking deeper and deeper. The *Galway Castle* sunk at 6:30 in the morning on September 15th and they headed back to Berehaven.

They arrived in Berehaven the next evening after getting off course in the fog a couple of times and left for Queenstown about midnight, arriving there at 6:00 the next morning. They and a British destroyer escorted a transport into Liverpool where the USS *Allen* then went into dry dock. Dry dock is a basin that can be flooded to let the ship float in and then drained to let the ship rest on a dry platform for repairs. Ralph wrote, "Went into wet basin about 10:30. I left the ship on a 7- day furlough."

Ralph had decided to spend his furlough in London and at 6:30 on September 19th he arrived there on the train and got a bed and breakfast for 7 shillings. The next morning, he was up at 8:00, had a breakfast of bacon, two slices of potato and tomato, and made his way to the American YMCA where he bought a sightseeing ticket for 4 shillings. The day was spent visiting the Tower of London, the Westminster Abbey, St Paul's cathedral and the House of Parliament, arriving back at the YMCA at 4:00 in the afternoon. He booked a room at the Bucks Hotel In The Strand where he stayed for at least two nights.

On the 22nd he walked around town in the forenoon and went out to Hydes Park to a kind of Wild West show in the afternoon. The next day he just bummed around at the YMCA. His last full day of furlough he took another sightseeing trip, went through the Buckingham Palace stables and saw King George's horse and carriages. He left London on the 2:10 train and was back on the ship at 7:30. Ralph was apparently not very impressed by the food in London, saying, "Was glad to get back where I could get something to eat."

Ralph makes no mention in his diary of the night life in London. The only clue I, Clarence, have is that several years ago in trying to decide what to get my dad for Christmas, I asked him if there were any songs that he would like to have a copy of and he said, "One More Time Around Piccadilly Circus." While I was never able to locate the song, it may indicate that Ralph had a good time visiting the shops, restaurants and bars of Piccadilly Circus in London.

His first assignment back on the ship was mess cooking for a few days followed by a not so favorable job working in the bilge under the boilers which would need to have the black oily residue from the boilers cleaned out from time to time.

Ralph reports in his diary on the unrest among the sailors on the H.M. Cruiser *Laviathion*. Apparently, the sailors had had a meeting and 50 of them walked off the ship in their work clothes and went to Liverpool followed a few hours later by 120 more jumping ship in their liberty blues. Ralph says that they were unhappy that their captain would not give them liberty so they just took it. Fifty armed marines were sent out after them and they returned that night either voluntarily or by force.

On October 10th they left dry dock, went to a tanker to take on oil and were under way for Plymouth, England getting there mid-afternoon the next day where they spent a day operating with American submarines and trying out a new submarine detector. Then they were off to Berehaven where they met three battleships and the next day met a convoy of several ships that they escorted into Liverpool. They left Liverpool for Milford Haven where they picked up a convoy which they escorted into Brest, France and then back to Queenstown on the 22nd.

They left Queenstown on October 28 and arrived at Liverpool about noon the next day. Ralph had liberty to go ashore at 2:00 and got back to the ship at 10:00. The next day they sailed out of Liverpool with the *Maritamia* but sprung a leak in the bow and had to turn back for repairs. With repairs done, they got to Holyhead at 4:00 in the afternoon. The next day they were back in Queenstown before noon. On November 2 they were given liberty from 1:00 in the afternoon until 9:00 the next morning.

On November 3, 1918 Ralph wrote in his diary, "About 8:30 this evening we got word that Austria had surrendered."

On November 4th they were assigned jobs around the ship and Ralph reported in his diary, "No boats are reported running on the surface flying a white flag. All the destroyers have been called in." For the next 5 days they worked on maintenance chores on the ship.

Saturday November 10th Ralph got liberty at 1:00 o'clock and he reported in his diary, "We got word this evening that the Kaiser has abdicated." The next day, Sunday, they had the whole day off.

Monday morning, they got word that the armistice was signed at one o'clock last night [early morning] and hostilities ceased at seven o'clock this morning. Ralph had the watch from 12 to 4.

The next morning, they went out expecting to have target practice but it turned out to be just a boat ride.

On November 14th they left Queenstown, met the *Mouritania* the next day and escorted her into Liverpool, and then returned to Queenstown.

On November 20th they left Queenstown for Berehaven where Ralph writes in his diary, "Monkeyed around in the bay outside Berehaven for the movie men." Then off to Queenstown they sailed and from there to Belfast,

Ireland. They arrived in Belfast on the 24th along with two other destroyers, the Cassin and the McCall. On the 27th they left Belfast for Holyhead, Wales. Ralph wrote in his diary, "We are on a pleasure trip." On the afternoon of the 28th the people of Holyhead gave a show at the Hippodrome for the Allied sailors. Ralph commented, "It was fine." He wrote to his sister Myrtle saying, "We got back to Queenstown Sat morning. Had a good time at Belfast and Holyhead." The next day they sailed back to Queenstown and on Monday December 2nd, Ralph took the examination for fireman first class and passed. He commented in his diary, "Will get $46 a month now." He probably would have been getting about $30 a month so this was a nice raise.

On December 3rd they went out for target practice but it was too foggy. Ralph recorded in his diary, "Six submarine chasers left for the United States today. When they sailed out all the ships in the harbor began whistling." A salute for a job well done. On the 5th the USS *Allen* left Queenstown for Brest, France and six more submarine chasers left for home with the same sendoff as the others. Arriving in Brest the next day Ralph got liberty at 2:00 and the next day liberty at 1:00.

December 8th was Sunday and Ralph once again commented on the meal they were served, "pork chops, corn, mashed potatoes, dressing, bread and fruit" followed by liberty so Ralph went ashore in Brest.

At 3:30 in the afternoon on December 12th they left the harbor in Brest, France on their way out to meet President Wilson. Ralph wrote his sister Myrtle saying,

> ... we are going out to meet President Wilson. There is going to be French, English and Italian battleships go to meet him besides the American battleships and destroyers. We are going to meet him about 600 miles out of France. That will be quite a trip ... I don't know whether I can get out when peace is signed or not. I am going to try to.

At about 9:30 the next day (December 13, 1918) they met the President's ship and escorted it into the harbor at Brest. At 3:30 that day they had a review of the destroyers with each one sailing single file past the president's ship, the George Washington, that was anchored in Brest harbor. Following the review, they left for Queenstown having target practice enroute the next morning. Back in Queenstown they were busy overhauling the boilers in preparation for sailing back to the United States.

President Wilson was off to Versailles near Paris where he would meet the heads of the other three of the "Big Four" nations and in the Hall of Mirrors they would hammer out the details of the Treaty of Versailles and the League of Nations. Russia was not involved in the negotiations because she under Lenin had signed a previous treaty with Germany and had withdrawn from the war. The big four were France, Italy, Britain and the United States with Italy dropping out before the conference concluded making it the big three.

France had suffered greatly as a result of the war, losing a fourth of its military age men and 400,000 civilians and having massive physical damage. They wanted safeguards against Germany. England had suffered a huge financial loss but not much physical damage and argued against revenge against Germany saying they would still have to live across the North Sea from Germany. Italy was not as hostile to Germany as the other three and dropped out of the conference negotiations early.

The United States under Wilson had 14 points that they wanted included in the final agreement but Wilson was willing to give many of the 14 up in order to preserve his League of Nations. In June of 1919 the Allies told Germany that if it didn't sign the treaty that the allies had agreed to, the war would resume. Five years after the assassination that started the war, the treaty was signed on June 28, 1919.

In the treaty Germany lost 25,000 sq miles of land and 7 million people. The treaty aimed to make Germany incapable of offensive war and required them to pay for the damage they had caused. President Wilson came away from the conference with the League of Nations agreed to and proclaimed that "at last the world knows America as the savior of the world."

But back in the United States the Republicans had taken control of the Senate in the 1918 election and were not impressed with the provisions of the League of Nations. It would require a two-thirds majority to ratify the treaty of which the League was a part and coupled with the fact that the President would suffer a debilitating stroke while on a speaking tour the summer of 1919, the treaty would never be ratified by the United States.

Christmas eve Ralph went ashore to the U.S. Naval Mens Club for a Christmas play. Christmas day he had the 8 to 12 watch, so had to stay on

Ralph eating pineapple in the Azores Islands.

board the ship but they had a big Christmas dinner with turkey, dressing, mashed potatoes, gravy, soup, asparagus, mince pie, fruit salad, bread and butter, peanut and chocolate candy followed by a cigar and a box of cigarettes.

On December 26, 1918 at 10:30 in the morning Ralph aboard the USS *Allen* sailed out of Queenstown harbor bound for home. Ralph recorded in his diary, "Everyone in Queenstown was out to see us leave. You can see heads and handkerchiefs waving from every window." This was a picture that Ralph would never forget and we heard him describe that scene many times.

Four days later they docked in the Azores Islands and were given 4:00 to 9:00 liberty. They were anxious to get home and had made 13 knots from Queenstown to the Azores Islands. The last day of the year they sailed out of the Azores Islands with four other destroyers who joined the USS Allen in shooting red and green skyrockets at midnight to celebrate the New Year.

The weather had been fine up until evening on January 4, 1919 when the sea got pretty rough but by evening on January 6th the weather had calmed and the USS *Allen* arrived at New York and anchored in the North River at 4:00 in the afternoon on January 7th. As we look at these

Storm at sea.

pictures of the ships tossing and turning in the angry seas we have to wonder how the sailors on board managed to do their daily jobs. Ralph's diddy box still contains most of the personal belongings that were in it during his navy years. One of those things is this razor. Imagine shaving with this razor when the room is constantly moving.

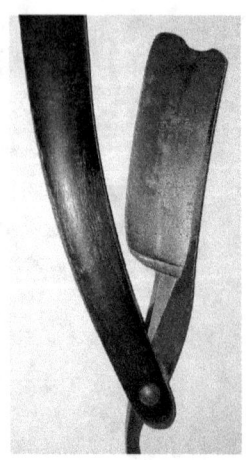

Ralph's straight edge razor.

Ralph went ashore in New York the evening they arrived there. He wrote his sister Myrtle saying, "New York is quite a city." On January 8th he left the ship on a fifteen-day furlough. He left the Grand Central Station in New York at 5:00 that afternoon and arrived at the LaSalle Street station in Chicago at 9:15 the next morning. He transferred to the Grand Central Station and left at 11:30 that night bound for Hampton, Iowa. When he woke up on January 10th· he was one station west of Dubuque, Iowa and arrived in Hampton at 10:30 that morning.

Brother Charles and his wife Ethel came to visit the next day and Velma came with them. The next day Ralph wrote, "Velma has got the flu."

The Spanish Flu pandemic, caused by an H1N1 virus, which would result in about 675,000 deaths in 1918 in the United States, and about 50 million worldwide, started in February 1918, about the time that Ralph was making his way across on the USS *Leviathon* enroute to Liverpool, England and the USS *Allen*. That crossing, we believe, was relatively free of sickness except for sea sickness, but by March the virus had reached military training facilities and the *Leviathon* would have a disastrous voyage enroute to Brest, France on September 29, 1918 when an infantry unit from Vermont boarded ship heavily infected by the virus. It is interesting to note that Assistant Secretary of the Navy, Franklin Delano Roosevelt, our future president, according to the *New York Times*, was removed

Ralph on furlough back in Hampton, Iowa, with little sis, Hazel.

from the *Leviathan* a few days before this fateful voyage, on his return trip from visiting the troops in France with what they termed the Spanish Grip. The USS *Allen* would have been a pretty safe place to be during 1918, we think, as long as you could avoid torpedoes and as long as no sailor on liberty brought the virus back to the ship. The government was attempting to censor the news about the flu but we think Ralph would have mentioned it had the flu arrived on the USS *Allen* while he was there.

In November of 1918 sister Myrtle had written from Julesburg, Colorado, where she was working, writing, "They closed everything here again Friday on account of the flu. There is several new cases. I guess there is several out in the country that has it. The Russians have it and they won't stay home, but go and give it to everybody.... Some of the folks at Iowa Falls have had the flu. They are having a teachers and board meeting here this afternoon to decide how long they will keep the schools closed. The board wants to close for a month and some of the teachers doesn't want to close for only two weeks. That is if there is no more new cases."

December 14, 1918 Ralph's brother Ray wrote saying, "I hope you don't get the flu. Earl and I had it. I wasn't so very sick but I was in bed for seven days and couldn't go out of the house for a long time. I was out quite a bit today and am awful weak. I don't think I can do much work for a week yet. I paid $350 for my Ford. I will send you a picture of it."

On the 14th Ralph went to visit his brother Earl, stayed overnight and the next day he and Brother Ray went back to Hampton.

On the 17th Ralph wrote in his diary, "Ray and I went out to Coulter on the ten o'clock train and Chas came to meet us in his Ford." The next day Chas and Ethel took them back to Hampton in the Ford.

The following day, the 19th, Ralph recorded, "Ray forgot to drain his Ford so he went back to Rockwell City today. He was afraid it would freeze up."

On January 20th Ralph boarded the train in Hampton bound for New York City. After transferring in Chicago he arrived back at Grand Central Station in New York City at about 3:00 in the afternoon on January 22. That night he stayed at the Soldiers and Sailors Club and got back to the USS Allen at 4:30 on January 23rd.

The war was over, Ralph had fulfilled his duty, and it was decision time. It had been about two weeks since he had gotten back from his furlough and on February 4, 1919 he talked to the executive officer and put in an affidavit to get out of the navy. Four days later he was notified to be ready to be discharged in three or four days. On the 10th Ralph went to Building 15 at the Brooklyn Navy Yard and took the physical examination for discharge.

On February 11th Ralph wrote in his diary,

> Left the USS *Allen* at nine o'clock this morning. Went to the destroyer pay office at 95 St., New York to get payed off. We had to wait until four thirty in the afternoon to get paid and didn't get back to the ship until six this evening. I lashed my bag and hammock together, got my discharge and other papers and left at eight o'clock for the Grand Central Station. Boarded train for Chicago at 9:17.

On February 13, 1919 he arrived back in Hampton. He had answered his country's call and now he could think about farming. Ralph had survived the First World War and would go on to a successful career as a farmer that he had wanted to do when he left Rockwell City for Montana on March 23, 1917. His ship the destroyer USS Allen would also continue on a successful career serving in World War II and surviving the Japanese attack on Pearl Harbor.

TO OUR SOLDIERS AND SAILORS.

God bless our boys who wear the blue!
God bless our soldiers, brave and true!
God bless our sailors, one and all,
Who answer to their country's call!
      God bless them all.

Here's to the boys who volunteer!
Here's to the boys who have no fear!
Here's to the boys who sail, who fly!
Here's to the boys who dare to die
      For freedom high!

Hurrah! Hurrah, for manhood brave!
Hurrah! Hurrah! let banners wave!
Let every mother sing with joy
  Who proudly gives a soldier boy!
Who proudly gives a hero grand
To guard our own beloved land!
Hurrah! Hurrah! defenders true
Under the réd and white and blue!
Here's to you, to you, to you!
      God bless you all!

—MARTHA S. GIELOW.

Thank you tribute poem to sailors and soldiers.

A F F I D A V I T.

State of Iowa,)
Franklin Co.,) SS.

I, A. H. Fitz, being duly sworn, do depose and say that I am the father of Ralph G. Fitz, First Class Fireman, U. S. Destroyer Allen, now stationed at New York Harbor.

Said Ralph G. Fitz owns a ranch of four hundred acres in Montana, and desires to be released from service so he can follow his occupation as an agriculturist and improve and crop said ranch. That during his time of service he has been unable to do anything toward improving his place and the same is now remaining idle, and that my health and age prevent me from being of any assistance.

Therefore, the undersigned respectfully requests that his son may be discharged from service as soon as possible in accordance with the judgment of the Bureau of Navigation.

Dated at Hampton, Iowa, this 29th day of January, 1919.

_____

Subscribed and sworn to before me by A. H. Fitz this 29th day of January 1919.

_____
Notary Public in and for
Franklin County, Iowa.

Request from his father for discharge from Navy.

CERTIFIED COPY OF SAILOR'S DISCHARGE   P6745 Form 104G

Form No. 82. Bureau of Navigation

United States Navy Discharge

THIS IS TO CERTIFY, That No. **10**   **RALPH GLENN FITZ**

a. **Fireman First Class,**   has this day been discharged from the

U. S. S. **ALLEN**   and from the U. S. Naval Service,

by reason of **SPECIAL DIRECTION OF THE SECRETARY OF THE NAVY. REFERENCE, BU., NAV. LETTER 225-18 (SERVED DURING WAR) Discharged at own request.**

Is ⸻ recommended for reenlistment. Rating best qualified to fill, **Fireman First Class**

Dated this **11th** day of **February** 19**19** at **Navy Yard, New York, N.Y.**

Character of discharge **ORDINARY**   **L. C. Farley, Commander** U.S.N.,

Commanding U. S. S. **Allen**

(SEAL)

Enlistment Record

SCALE OF MARKS: 0, Bad; 1, Indifferent; 2, Fair; 2.5, Passing; 3, Good; 3.5, Very Good; 4.0 Excellent.

Name **RALPH GLENN FITZ.**   ; Rate **F. 3 c**

(AT ENLISTMENT)

Enlisted **July 12th** 1917, at **Salt Lake City, Utah** for **4** years;

Previous Naval Service – – – – years; Served Apprenticeship – – – ; Gun Captain Certificate – – –

Certificate Graduation P. O. School – – – – ; Seaman Gunner – – – – ; Trade **None**

Citizenship **U.S.** ; Ratings held during enlistment **F 3c; F. 2 c; F l c**

Proficiency in Rating **3.5** ; Seamanship – – – – ; Ordnance – – – –

Signaling – – – – ; Marksmanship, Small Arms – – – ; Mechanical ability **3.4**

Knowledge of marine machinery **3.4** ; Knowledge of electrical machinery and appliances – – –

Sobriety **4.0** ; Obedience **4.0** ; Average standing for term of enlistment **3.6**

W. W. Webb Lieutenant U.S.N.

and Executive Officer

Descriptive List

(To be made after careful examination at date of discharge.)

Where Born **Churdan Iowa.**   Date **June 5,** 1**894**

Age **24** years **8** months; Height **5** feet **10½** inches; Weight **160** lbs.; Eyes **blue**; Hair **brown**; Complexion **Ruddy**

Personal characteristics, marks, etc. **S. L Index finger P.M. r Scapula**

Percentage of time on sick list during enlistment **.0%** Is – – – physically qualified for reenlistment.

F.J.D. Breen Lt(jg) M.C. U.S.N.

(SIGNATURE OF MEDICAL OFFICER)

I hereby certify that the above named **Ralph Glenn Fitz**

has been paid **Twelve – – – – – – – – –** Dollars **Ninety** Cents

($ **12.90/100**), in full to date.

Trans. waived

No mileage paid   **February 11** 19**19**

W. A. Cally Lt. P.C. U.S.N.

(SIGNATURE PAYMASTER)

RF

Filed for Record this **8** day of **April** 19**19**

Ralph's Discharge.

CHAPTER 7

RE-ENTERING CIVILIAN LIFE

R alph returned to Hampton, Iowa from service in the Navy during World War I on February 13, 1919. While he probably was thinking about his homestead in Montana, it was winter in Montana and besides, he had family to catch up with after having been away for almost two years except for one short furlough.

Fitting bits and pieces together it seems that Ralph probably had his neighboring homesteader Gust Pehlke farming his land while he was serving in the Navy. So when he returned to Montana in the spring of 1919, Ralph got a job working for Bill Scheele and through exchanging help with haying, Ralph and Reinhard Sudbrack became

Gus Pehlke's house.

Gus Pehlke's farm.

Making hay in Montana

friends. Reinhard wrote in his memoirs that he and Ralph got along very well, enjoyed each others company and ran around together at nights. It was probably while working for Bill Scheele that one of Ralph's favorite stories occurred. Most likely he ate meals with the Scheele family and Ralph

Reinhard Sudbrack's farm.

Reinhard Sudbrack's house.

told the story of having to watch for flies in the fried potatoes. Having visited that area of Montana in late summer, I, Clarence, can vouch for the unbelievable number of bugs that were impossible to keep out of the house. The window sills were full of bugs, the light fixtures were full of bugs, and while outside my body was covered with what looked like giant mosquitoes.

Reinhard said he and Ralph ran around together at nights but he gave us no explanation as to what they were running around doing at nights. The war was over and the Roaring 20's were creeping in, evidenced primarily by women with short skirts revealing their knees, by their hair cut in a bobbed style, by their dancing, smoking and drinking in public and their freedom to mingle with men. We got a clue from a letter that Reinhard wrote to Ralph that they went to a lot of dances. Much fun one might think for men in their twenties like Ralph and Reinhard, but they are mum on the subject.

The 18th amendment to the U.S. Constitution, the amendment that ushered in prohibition, was passed by Congress in December of 1917 at the same time that Ralph was being processed for active duty in the Navy. By the time it would be ratified by two-thirds of the states, Ralph would be on furlough back in Iowa and by the time prohibition went into effect Ralph would be back in Montana.

Montana was officially on the "dry" side of the prohibition issue with many of the eastern states being "wet" states. But as the twenties progressed, the speakeasies arrived in Montana, at least in the larger cities like Billings. Undoubtedly not in Whitewater. Ralph made no mention of alcoholic beverages in any of the things he wrote and we suspect that he was officially on the "dry" side. During our lifetimes I, Clarence, never saw Ralph drink an alcoholic drink. My co-author, Mardelle, as a young girl, relates that she thinks she saw Dad Ralph drink a beer with Reinhard Sudbrack when the Sudbracks visited in Iowa. Does that give us a clue as to what Reinhard and Ralph were doing when they ran around together at night? Probably not in Whitewater but maybe in Saco or Malta and Ralph had a car. Only once did I, Clarence, see him smoke a cigarette. Dad and I stopped to visit Russell (Jack) Lindsey and as we sat in the car visiting, Jack offered Dad a cigarette, but as he took it he said he probably shouldn't do this anymore because of the boy in the back seat. The boy in the back seat wasn't supposed to be listening, but he was.

In wintertime, when work on farms was scarce, the timber camps became active, that being the best time for lumbering. The Homestead Act had a requirement that you live on the homestead but Ralph was able to get a leave of absence from the Department of Interior to be away from his homestead from August 25, 1919 to December 31, 1919. So, in November of 1919, Ralph was working in a timber camp in Idaho and probably living in the town of Fortine, Montana located in the northwest corner of Montana and close to the Idaho border or perhaps he lived at the lumber camp and just mailed a letter from Fortine.

Timber camp, Idaho.

Timber camp.

Wheat harvest and wheat shocks.

As the fall of 1920 arrived, and the wheat was harvested in Montana, Ralph and Reinhard decided to try working the wheat harvest in Canada. Reinhard wrote in his memoirs, "When we took off for Canada, we were told at Whitewater that all you had to do when you crossed the line was to report to the nearest custom office, (in our case Pontiac). There we were told that we were already in error, but the custom officer told us that since we were only planning to stay 3 weeks, we'd be all right as long as we didn't drive a car. Ralph took his car to a garage to have the valves ground and kept it there during his stay in Canada. He had a very nice bed roll, complete with good woolen blankets and quilts, covered with canvas. Our first night in Canada, it snowed 2–3 inches, and I tell you it was pretty tough to crawl out of our nice warm bed to put on our boots in the snow. Fitz was an ambitious fellow and during bad weather when we couldn't thresh, he hired out

Entering Canada with Ralph's car.

picking rock at $6 per acre. When the threshing was over, Fitz returned to the United States in the allotted time but I took off and worked in a lumber camp for the winter."

Ralph apparently went back to Iowa sometime after he got back from working in Canada but we have no record of when. Maybe he continued to work for Bill Scheele but we have no evidence of his whereabouts until he mailed a letter from Hampton, Iowa on August 26, 1921.
On September 1, 1921, Gust Pehlke wrote Ralph a letter:

> Dear Ralph,
>
> I have not heard from you for a long time. Expected you here for threshing. We threshed August 28. Got 171 bushels wheat on your place and 149 bushels oats. Wheat around here runs 5 to 12 bushels per acre. South of Saco 26 to 35. Around Greve [a small community that no longer exists] nothing.
>
> The wheat had a very good start this spring but we had a dry spell later part of June. First part of July we had a three day rain but it was too late to save the wheat. Couple weeks later we got a hail storm that finished it. It started near Wise and took in Ela Simonsen, Irene Erickson, yours and mine, part of T. Olson, part of L. Christopherson and B. Trary, S. Seimonson place, part of Elmer Erickson and Rushman place and stopped at Stinky. This was about 5 o'clock in the morning. At 5 o'clock in the evening it hailed north and took everything. I got 329 bushels from 65 acres. The oats I cut

for hay. It turned out better than I expected. No insurance. Corn was stripped. Just the stem left but it picked up again and is getting ripe.

I took one load Wheat from your place to Saco. It weighs 58 pounds to the bushel with 1-1/2 pounds docked. At that rate we will have 161 bushels on your place. Oats is 15 cents a bushel now. I got 1.06. Wheat that weighs 59 to the bushel was 1.10. 61 to the bushel is 1.20.

Let me know if you want your wheat hauled and stored. I got it separate in the center bin and like to get it out to get the fanning mill back in the granary and we can tell better how many bushels we got on your place by hauling it in first. The oats you may leave til later.

I plowed the 10 acres where I had the oats on. Everything is the same as always only very hot. Good for corn. To dry for plowing. Seimonson got the threshing around here …

Best regards from all of us
Gust Pehlke

This letter from Gustav Pehlke was addressed to Ralph in Hampton but then forwarded to him at Hubbard, Iowa. We don't know what he was doing in the Hubbard area, but he was presumably working at some job. It is interesting to note that at about that same time, Ralph had a savings account at the Farmers National Bank of Oskaloosa. Might they have had a branch in Hubbard? Or was he traveling to wherever the work was available. If he was having second thoughts about farming in northern Montana, this letter from Gust would not have given him any encouragement.

And then to increase his doubts, in a letter from Reinhard in the fall of 1921, Reinhard wrote that the crops this year were about half normal, about like last year. We have no evidence that Ralph spent any significant amount of time in Montana after that but we have scant evidence that he didn't either.

We think that the homestead residence requirement on his homestead had been fulfilled and on May 19, 1921 President Warren G. Harding awarded a patent on his 400 acre homestead to Ralph. He was now free to live and work wherever he wanted.

Economic times were good following the end of World War I and through much of the 1920's. Herbert Hoover who would become the 31st president of the United States said during his campaign, "We in America are nearer to the final triumph over poverty than ever before in the history of any land ..."

At his inauguration on March 4, 1929, President Hoover stated, "Ours is a land rich in resources ... filled with millions of happy homes; blessed with comfort and opportunity. ... I have no fears for the future of our country. It is bright with hope."

On October 29, 1929 the headline of the *New York Times* read:

Stock Prices Slump $14,000,000,000 in Nationwide Stampede to Unload ..."

The Great Depression had begun.

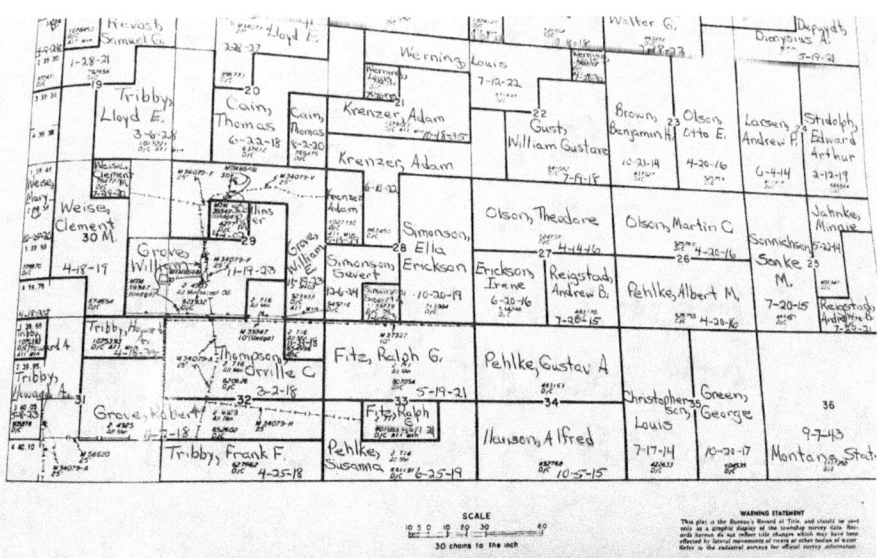

Platt map, Montana homestead, Sec 33, 1921.

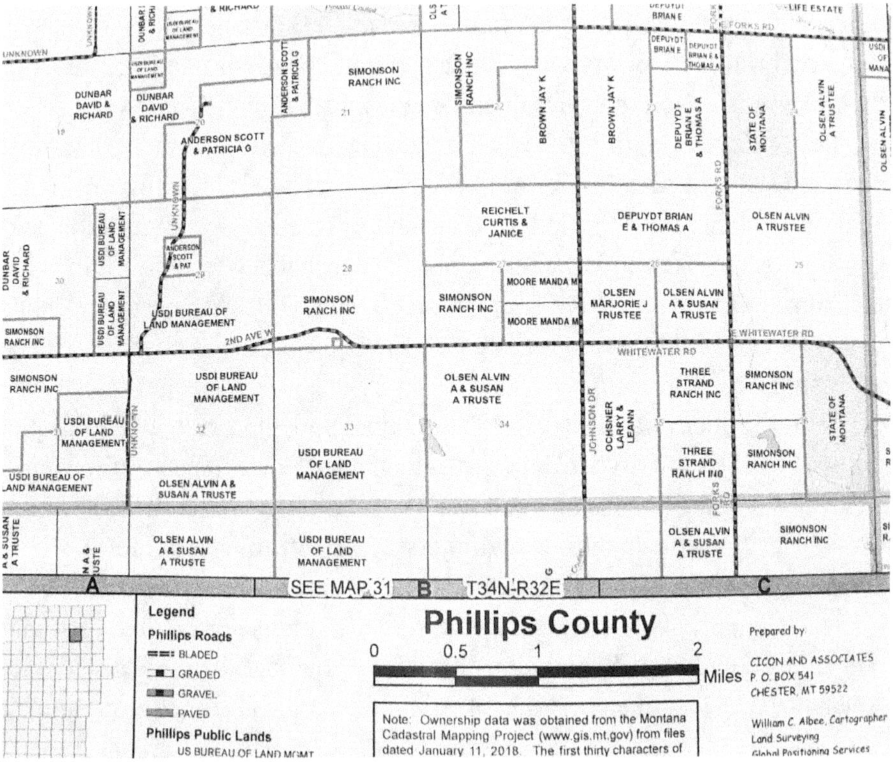

Platt map, Section 33, BLM land, 2021.

CHAPTER 8

THE FARM WHERE WE, THE AUTHORS OF THIS BOOK, WERE BORN

The south half of the land which would become the home of the Ralph Fitz family first came into the Fitz family when it was purchased by our Grandma Addie Fitz in 1910. We think she probably used her inheritance to buy the farm; her father died in 1905 and her mother in 1908. But why Franklin County when she and Grandpa Alvin had lived most of their lives and raised their family in Greene and Calhoun Counties?

Editor Raymond, in 1907, wrote an article in which he opined that most people came to Franklin County because someone else, family or friend, already lived there. We think that was probably true for Grandma Addie and Grandpa Alvin. We think that attraction was probably Grandpa Alvin's oldest brother, Great Uncle John Abraham Fitz , who while we have no evidence that he owned land there, was listed on the 1900 census as a farmer in Lee Township, Franklin County.

John Abraham was born in 1836 in Ohio before the family had migrated to Iowa. In 1861 the Civil War began, and at age 25 he enlisted in the Union Army, Iowa being a free state, and was assigned to Crocker's Brigade which was a brigade made up of four regiments, all infantrymen from Iowa.

John Abraham served four years and took part in some seventeen heavy battles and many small ones. He was part of Sherman's march to the sea that was a "scorched earth" campaign that broke the back of the Confederate resistance. He was taken prisoner at Atlanta, Georgia and spent four almost

Andersonville prison – Photo taken by A.J. Riddle on Aug. 17, 1864. Reprinted from The Prison Camp at Andersonville, *a National Park Civil War Series.*

intolerable months in Andersonville prison followed by three additional months in Libby prison where conditions were noticeably better. He was released from Confederate prison at the end of the war which ended May 9, 1865 and married Caroline Upmeier that December. They had several children some of whom made their homes in Franklin County.

In the book, *History of Franklin County, Iowa, a Record of Settlement, Organization, Progress and Achievement*, edited by I.L. Stuart, and published in 1914, some interesting facts and opinions are recorded.

The following are excerpts from a chapter written by attorney B.H. Mallory, a member of the Franklin County Bar:

> I shall discuss the subject of drainage solely as it applies to the reclamation in whole or in part of lands too wet for food-producing purposes. Man has ever been prone to think he could improve upon the works of the Almighty in creating the world and especially has this been true with the

relation to the distribution of water over its surface and in many instances the hand of man has supplemented the hand of God to man's profit and God's glory.

Franklin County, on account of its geological formation rather than on account of its elevations, in its natural state, contained a very large percentage of wet land.

With the exception of a small area which drains to the south and west towards the Iowa river, the drainage of this county is south and east through various creeks into the Cedar river. ... if you should drive north and south through the county on the section line running through the city of Hampton, you would cross a large number of beautiful creeks, some of them quite large. If you would parallel this road four miles west, you would find that these creeks had lost themselves and that the lands in that portion of the county originally were without outlet, except for extreme overflows, and a large portion of the same consisted of sloughs, ponds and marshes.

The first drainage district was established October 3, 1905 ... Since that time about thirty drainage districts have been established in the county, all but one of them in what is known as the western part of the county. The general plan has been to dig dredge ditches emptying into the headwaters of the present creeks and by this method the creeks, the natural drains of the county, have had their lengths extended largely to the west. These open ditches have formed the outlets for tileage systems ...

This system put in by and paid for by the county, then made it possible for individual owners, at their own expense, to tile their land and run the excess water into the county system and in turn into the natural drainage system.

If you look at an Iowa map you will see that the area where Grandma Addie's land and which eventually would become our Dad Ralph's land was located would have benefited greatly by the drainage system Mr. Mallory describes. Did Grandpa Alvin and Grandma Addie look into the future and see what this land could become? Maybe? Or was it luck. But even with tile having been installed to direct the water to the drainage ditch, we would sometimes get standing water after a rain that would drown or stunt the growing crops.

Just to the north of our homestead was a field with a substantial peat bed that extended a way into the field to the west and on one occasion, I Clarence, remember a heavy rain that overwhelmed the drainage system. Because the milk cows were pastured in the field to the west, we had an electric fence attached to the top of the permanent fence on the fence line to keep the cows in their pasture.

On that particular occasion the rainwater was standing up to the top of that fence. Because we didn't want the electric fence to short out and become nonfunctional, it was my job as a high school age boy to get into my swim suit, wade into the water that was probably four or five feet deep and raise the electric fence above the water. But on years with less rainfall the field with the peat bed raised abundant yields.

I, Clarence, though can't let this subject go without mentioning how I hated tilling that field. Once the peat reached my skin I would start to itch and I remember one occasion where I finished the tilling as soon as possible and went immediately to the house to shower and remove the peat and the itching and get into clean clothes.

> Scott township was one of the last to be organized in Franklin County," according to *History of Franklin County*. "Prior to 1878 it was part of Morgan Township... Scott lies in the extreme western part of the county ...The land is of gently undulating character and practically the entire area is tillable prairie land, producing splendid crops yearly ... Lands reclaimed by these large ditches [Drainage Districts] have become very productive and their market value increased accordingly. Up to the latter '80s Scott was comparatively a new township, but little developed, with large tracts of choice land held by speculators.

The first settler in Scott township was a man by the name of Malin, who came in 1860 but only stayed two years. The next was John Martin who was employed to manage the Cobb farm which had been entered by Mr. Cobb in 1859. The Cobb farm contained something over five thousand acres and large herds of cattle and horses were raised on this farm for market.

Arrival of the railroad lines pretty much dictated where the towns would be. Hampton was an exception, with a community starting there in 1855. The first railroad to arrive in Hampton was the Central of Iowa in 1870. E.L. Cook is claimed to be the first Franklin County resident in the mid-1850s. Cook located in Maysville, located 6 miles south of Hampton, a town that had hoped to become the county seat, but when the railroad arrived in Hampton instead of Maysville, Hampton became the county seat and Maysville eventually disappeared. In 1881 the Central Railroad of Iowa arrived in Alexander and the town was platted in 1885. Latimer was established in 1882 and the Iowa Central Railroad had a station there. Coulter grew up around the Chicago and Great Western station established there sometime in the 1880s.

The area was developing rapidly when Grandma Addie Fitz purchased the south half of the farm from J.L. Sinclair and his wife, Mabel C. Sinclair for twelve thousand four hundred two dollars, described as "the South East Quarter (SE 1/4) of Section 28 (28) Township Ninety Two (92) Range twenty two (22) West of the 5th P.M. Iowa. Except one acre commencing at the south east corner of said land running thence North 16 rods, thence West 10 rods, thence south 16 rods, thence East 10 rods to place of beginning. Subject to a mortgage of Four Thousand Dollars due in four years from March 1st 1910, drawing 5% which the said Addie E. Fitz assumes and agrees to pay."

On March 27, 1925, Fred Nancolas and his wife Ida Nancolas sold the north half of the farm described as the "Northeast Quarter (NE ¼) of Section Twenty-eight (28), in Township Ninety-two (92), Range Twenty-two (22), West of the (5th) Principal Meridian, containing One Hundred Sixty (160) acres more or less, according to the United States Government Survey thereof" to O.F. Donaldson of Hamilton County, Iowa, and on that same day O.F. Donaldson and his wife, Ella C. Donaldson sold the same property to the Jefferson Land Company of Winona, Minnesota.

On August 31, 1927, the Jefferson Land Company leased to Herman Varrelman the Northeast quarter it had purchased two years earlier for a term of three years (March 1, 1928 to February 28, 1931) on a combination crop share-rent lease subject to the sale of the property. For whatever reason, unknown to we the authors, the lease was reassigned to Ralf G. Fritz (sic). Perhaps it was a sub-lease because Herman Varrelman still appeared to have an interest in the lease at the time the property was sold.

A note of explanation: our research indicates that O.F. Donaldson was an agent for the Jefferson Land Company. Was the Jefferson Land Company owned by O.F. Donaldson? Perhaps. And Frances Donaldson who we think was subordinate to O.F. Donaldson in the company may have been a daughter or some other relative.

On October 15, 1929, the Jefferson Land Company sold the property just described, to Frances Donaldson of Hamilton County, Iowa. We do not know the relationship between O.F. Donaldson and his wife Ella with Frances Donaldson nor do we know their relationship or dealings with the Jefferson Land Company and the legal records do not give us any clue. What we do know is that this transaction took place at the same time as the United States economy was entering a depression that would last until World War II started, a period of about twelve years.

We have not found much information on Ralph's life for those years between his discharge from the navy and his renting the north 160 acres of the farm that he would eventually buy. That period ran from his discharge in February 1919 to March 1928, a period of nine years. It appears that after spending a few weeks getting reacquainted with family and adjusting back to civilian life he went back to Montana until sometime in 1921. During that period it appears that he worked various jobs in Montana, Idaho and Canada so we suspect he was doing something similar once back in Iowa in 1921. Was he courting his future wife, Edith? Maybe.

His homestead in Montana, it was becoming evident, was not the "good piece of land," that he had told his father that he thought he had bought. The 1921 crop was dismal with a total income of about $200. We have no evidence of what the expenses were and what his arrangement was with Gus Pehlke, but we know the taxes were about $120 that year. Whatever the circumstances were, Ralph did not get the 1921 real estate taxes paid and the 1928 taxes were also due, so by July of 1929, Ralph received notice that his homestead was being sold to satisfy the taxes due. Again we have not found the details but Ralph obviously was able to satisfy the demand for taxes, because he was able to sell the property in 1930.

While the legal records are about as clear as mud, we are convinced that the depression had a lot to do with these rapid-fire transactions. At least I, Clarence, had been under the impression that our father, Ralph, had traded his homestead land in Montana for the farm in Iowa. Perhaps that was not technically correct but here is what we could find in legal documents.

On March 17, 1930 O.F. Donaldson and his wife Ella C. Donaldson sold to Ralph G. Fitz, "The Northeast Quarter (NE 1/4) of Section No. Twenty-eight (28), in Township No. Ninety-two (92) North, Range No. Twenty-two (22) West of the 5th P.M., Iowa, with the appurtenances thereunto belonging containing -- acres more or less according to United States survey, and do hereby covenant to Warrant and Defend the title to said Ralph G. Fitz against lawful claims of all persons except a mortgage of $14,000 with interest from March 1, 1930, which grantee assumes and agrees to pay." Ralph could now continue farming the land that he had been leasing since March 1, 1928.

So Ralph agreed to assume and pay the mortgage of $14,000 that the Donaldson's owed on the property. Did he do that by deeding his 400-acre homestead in Montana to the Donaldson's? We think so because on April 10, 1930,

> Ralph G. Fitz, single of Franklin County, Iowa, in consideration of the sum of One Dollar and other valuable considerations in hand paid, do __ hereby SELL AND CONVEY unto O.F. Donaldson of Webster City, of Franklin County, Iowa, the following described premises situated in Phillips County, State of Montana, to-wit:
>
> The North Half and the Northeast Quarter of the Southeast Quarter and the Northeast Quarter of the Southwest Quarter of Section 33, Township 35, North Range 32, East of the Montana Meridian, with the appurtenances thereunto belonging; containing 400 acres more or less according to United States survey ....

Was it a trade? For all practical purposes, we think so.

As we wrote earlier in this chapter, Grandma Addie purchased the south half of the half section farm in 1910. We believe Uncle Charley C. Fitz lived there and farmed the land from 1913 until sometime between Grandma Addie's death in 1925 and probably 1934. Uncle Charley married Ethel Hooker in 1915 and to them were born two daughters, one of which died as an infant. Charley and Ethel adopted a daughter, Alvena, who had a special relationship with the family farm that we will discuss later.

Grandma Addie had apparently willed her interest in the south half of the farm to her children and on August 29, 1934 Uncle Charley, now a widower, sold his interest to his sister Aunt Myrtle for $350.

That same year, on November 16th, Uncle Earl and his wife Alberta, still living in Calhoun County, sold their interest to Ralph for Nine Hundred Ten Dollars. Only a few months earlier on March 14, 1934 Ralph had married Edith Hill and in less than a year one of the authors of this book, Clarence, was born.

Also in that year, on March 1, 1935, Ralph and Edith mortgaged the northeast half of the farm to a neighbor, Ed Wilkins, in return for a $14,000 loan to be repaid in five years. It might be easy to confuse this loan with the purchase of the northeast quarter from O.F. Donaldson because the amounts are the same but the timing is wrong. Then on January 19, 1937 Ralph purchased the rest of the southeast quarter from his dad Alvin Mindon, his sister Myrtle, his sister Hazel and her husband Ralph Marken, his sister Velma and her husband Hiram Heideman and his brother Ray and his wife Mildred.

One has to wonder how much the depression was playing on the financial decisions the family was making. Perhaps Ralph had an advantage over his brothers and sisters simply because of timing and the events of the day. He had spent two years in the navy for which he was paid with very little place to spend it except on an occasional liberty.

He was very frugal and we would suspect that he sent most of his pay home to his father for safekeeping. In addition, not long after he returned home from the navy, he had 400 acres of Montana land free and clear. So he was in a good position to make the financial decisions that he did. Not that there weren't difficulties, the economy being what it was, because there were. But he was a hard worker, he was frugal, he was honest and he was single, all attributes that helped him survive the hard times

Uncle Charley rented and farmed the south 160 acres from his mother Grandma Addie from about 1913 to 1934 when, Grandma having passed

arley rented a house & land located North of
...Mildred's home Latimer IA. During that...

Uncle Charlie and Aunt Ethel renting south quarter section.

away in 1925, Ralph purchased the 160 acres from his Dad Grandpa Alvin and the heirs who still maintained an interest. A notation on a picture of Uncle Charley and Aunt Ethel in front of the house on the south 160 acres where your authors were born, indicates that Uncle Charley ran a gas station on Highway 1 which is located in the southeast part of the state and runs from Anamosa, Iowa to Keosauqua, Iowa. (That seems questionable. Perhaps the Highway numbers have changed?) We have no further knowledge of that.

I, Clarence, have a faint memory of visiting Uncle Charley at a house on the southwest corner of the section, about a half mile from our home. If my memory is correct, then that would have been sometime between 1935 and 1939 which is when Uncle Charley moved to the farm he had purchased east of Dows. I would have been between 4 and 5 years old. Uncle Charley and Aunt Ethel were parents to two daughters, Mary Catherine and her sister who died in infancy. Charley and Ethel completed their family by adopting Alvena, who would once again live with her husband Sydney Early, on the land where her parents lived when she joined their family. Sydney worked for Ralph for several years.

CHAPTER 9

PURSUING THE DREAM

It was 1934 and with a team of horses, control of 320 acres of farmland and a dream, Ralph set out to develop his half section of land into a productive economic unit. We assume that Ralph had lived in the house located on the north quarter section while he was leasing that quarter. We know that the house was habitable because after Ralph had purchased the north quarter it was rented to someone at least for a while.

The old house.

Eventually it fell into disrepair and was torn down. The only other useable building that I, Clarence, can remember is a hog house that was occasionally used to house pigs as they grew to market weight. Again we assume that after Ralph had purchased the south quarter and Uncle Charley and Aunt Ethel had moved out, he moved to that house. It wasn't a very fancy house. Probably wouldn't even be considered habitable by 2022 standards. Some years ago I, Clarence, sat at my desk and described that house as I remembered it. Here is what I wrote:

The house where I was born I remember as being gray. It was a wood frame house with wood lap siding. I don't know if it was gray because it hadn't been painted for so long or if it was cedar siding and was left to age and gray on purpose. It was a two-story house with an open porch on the south side that started at the southwest corner and went about halfway across the south side. There were a couple of steps from ground level to the porch floor. The ice box sat on the porch to the left of the door that led into the kitchen. The ice box was wood with an upper door on top in which a block of ice could be placed. The bottom door was then refrigerated and for food storage.

Inside the kitchen door on the south wall was the telephone. It was a wooden wall phone about 24 inches tall. There was a crank on the right side as well as a forked hanger that the receiver or ear piece hung on. The weight of the receiver on the forked hanger took the phone out of operation. To operate the phone, you would remove the receiver, and turn the crank on the side of the phone. This would signal central, as she was called, at the switchboard to ask what number you wished to call. She then would connect you with that number and you would talk to them through a bell-shaped mouthpiece on the front of the phone. But don't say anything that you don't want the whole neighborhood to know because anyone who had the receiver to their ear on your party line could listen to your conversation.

On the west wall of the kitchen was a large cast iron cook stove. The fire box was on the left and the oven on the right. As I remember, we used a lot of corn cobs, that were a byproduct of shelling corn for market, as fuel in the kitchen stove.

I can only remember two rooms on first floor. The living room was on the east side of the house. There was a kerosene burning stove on the north side of the room. It was a brown colored heating stove with isinglass windows on the front. There was a large ceiling register that allowed heat to go through the ceiling to the upstairs.

I have no recollection of the upstairs or where the stairway was to get upstairs.

This is the house where I was born and where I lived for the first seven or so years of my life. We called it "the old house."

It wasn't much of a house, but all Ralph needed was a comfortable bed and a place to eat a meal because he was spending long hours working on his farm. It took a long time to plow a field with a team of horses and a one-bottom plow. Now in order to make it a home, he needed a wife. He was ready. We don't know where he might have met his future wife or any other also-rans. We are pretty sure he didn't frequent the saloons.

Ralph accepts Jesus Christ in 1913.

Was it a blind date set up by a relative or friend? Perhaps it was at church, since Ralph was a regular church goer and had accepted Jesus Christ as his personal savior at age 19.

Did he buy Edith's fried chicken at a Box Social? A box social was an occasion where young ladies would make a box lunch which would be auctioned off and the young man, or anyone for that matter, who won the bid was able to enjoy the lunch with the lady who prepared it.

Edith was an attractive young woman at age 25, and Ralph certainly might have enjoyed having a box picnic with her. That would have been 1928 and Ralph was just get-

Edith.

Probably Ralph's car.

ting started farming on the north quarter section. We don't know. But if that were the case, they could have gone for a ride in Ralph's car on a Sunday afternoon. We think Ralph went to church at the Rowan Methodist Church. We know that Edith and her parents, Grandpa and Grandma Hill, also went to the Rowan Methodist Church. We know that Edith worked for Agnes and Raymond Hansen in some capacity for at least one summer, for which she was paid on one occasion with a five-dollar gold piece, which she treasured and kept throughout her life.

Agnes and Raymond lived about halfway between Ralph's leased farm and the Rowan Methodist Church. We know that Agnes and Raymond gave Ralph and Edith a framed picture as a wedding gift that hung over the couch (we called it the davenport) for as long as I can remember. Is there any connection? We don't know. Conjecture at best. The only thing I, Clarence, ever heard about Ralph's courtship of Edith involved another man that

Wedding gift.

Edith had gone riding with, and they tipped the car over in the ditch. And because of the accident, Ralph found out about it. Oops. Ralph would occasionally tease about that Sunday afternoon.

Whether it was a long courtship or a short one we don't know, but wedding announcements were sent and on March 14, 1934 Ralph married Edith Aleta Hill in a double wedding at Edith's parent's home near Dows, Iowa, along with her sis-

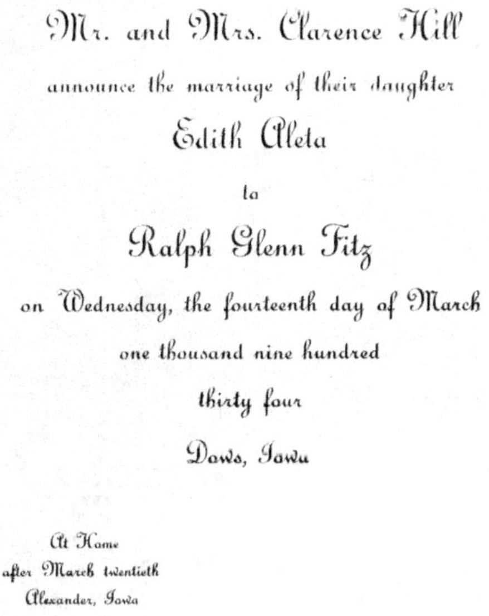

Mr. and Mrs. Clarence Hill

announce the marriage of their daughter

Edith Aleta

to

Ralph Glenn Fitz

on Wednesday, the fourteenth day of March

one thousand nine hundred

thirty four

Dows, Iowa

At Home
after March twentieth
Alexander, Iowa

Wedding announcement.

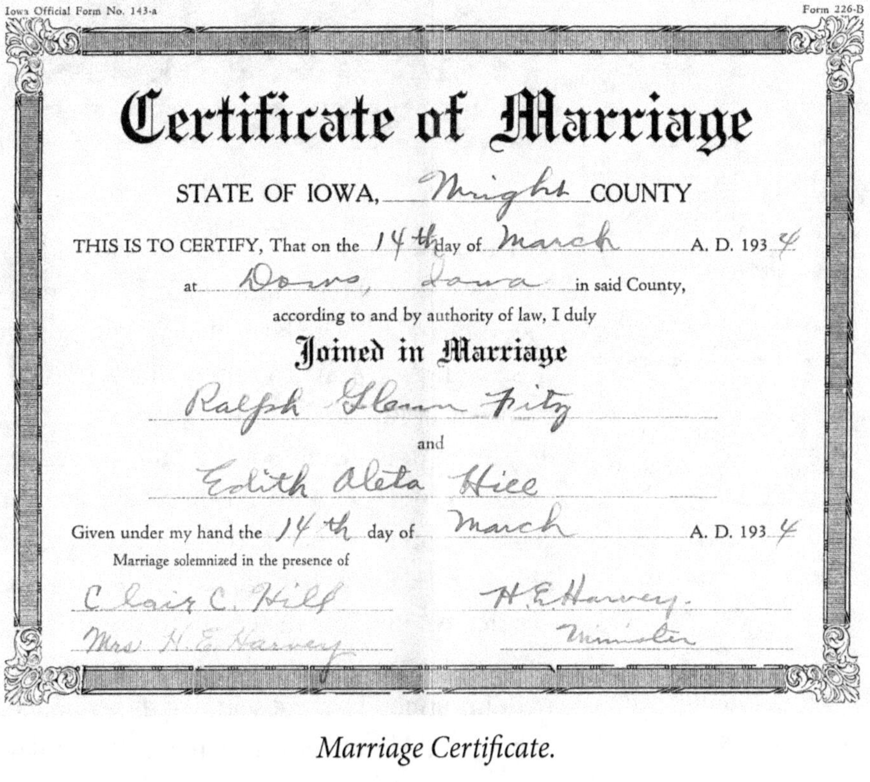

Iowa Official Form No. 143-a   Form 226-B

Certificate of Marriage

STATE OF IOWA, _Wright_ COUNTY

THIS IS TO CERTIFY, That on the _14th_ day of _March_ A. D. 193_4_ at _Dows, Iowa_ in said County, according to and by authority of law, I duly

Joined in Marriage

Ralph Glenn Fitz

and

Edith Aleta Hill

Given under my hand the _14th_ day of _March_ A. D. 193_4_

Marriage solemnized in the presence of

Clair C. Hill   _H. E. Harvey._

Mrs. H. E. Harvey   _Minister_

Marriage Certificate.

C. Hill Home Scene Of Double Wedding

Miss Edith Wed Ralph Fitz And

LaVonne Wed Dwaine Morris

Wednesday Afternoon

A very pretty and unusual event occurred at the home of Mr. and Mrs. Clarence Hill Wednesday afternoon at two o'clock, when their daughters the Misses Edith Aleta and LaVonne Esther were united in marriage to Ralph Glenn Fitz and Ollie Dwaine Morris, respectively in a double wedding ceremony. The room was tastefully decorated, the color scheme being pink and white.

To the beautiful strains of Lohengrin's Bridal Chorus, played on the piano by the bride's sister Miss Evelyn, the young couples took their places, unattended under a pretty arch and the Reverand H. E. Harvey, pastor of the Rowan M. E. church, read the impressive single ring service in the presence of the immediate families of the contracting parties.

The brides were attired in gowns of Baby Blue Crepe and the grooms were dark suits.

Ice cream and cake were served after congratulations had been extended the young couples and they left soon after by automobiles for short wedding trips.

Both young ladies were born and grew to young womanhood in this vicinity where they have many friends. Mrs. Fitz, the oldest daughter of Mr. and Mrs. Hill has assisted her mother at home since graduating from the eighth grade. Mrs. Morris attended the Dows high school and graduated with the class of 1931. Since then she has also assisted at home.

Ralph Fitz is the youngest son of Mr. R. G. Fitz of Hampton. He is an ex-service man, having served in the navy during the World War. Since then he has been farming near Alexander where they will be at home to their friends after March 20.

Mr. Morris attended the Belmond schools from which he graduated with the class of 1929. Since that time he has been assisting his father on his farm near that town, where they will also be at home to their friends after March 20.

The Reporter joins their many friends in wishing both young couple a happy and prosperous journey through life together.

Wedding announcement from the newspaper.

Wedding picture.

ter LaVonne who married Duane (Pete) Morris. Edith was one of ten children born to Clarence and Mildred Hill who farmed in Wright County close to Dows. Aunt LaVonne and Uncle Pete farmed near Rowan for several years after which they moved to Minnesota. Ralph, in relaxed moments, and with a twinkle in his eye, enjoyed giving his opinion that he thought the preacher, Rev. H.E. Harvey, had gotten mixed up and married he and Pete instead of to the girls. We have no knowledge of a honeymoon but their wedding announcement said that they would be at home in Alexander on March 20th, six days after the wedding. I, Clarence, have often thought that it must have been true love for Edith to move into Ralph's house. There was no running water, therefore no water heater except for a pot on the stove. The toilet was an outdoor traditional outhouse with the traditional Sears Roebuck catalog for tissue. Bathing took place on Saturday night in preparation for Sunday School and Church on Sunday morning and was in a metal tub set on the floor and filled to an adequate level with water warmed on the kitchen stove.

In spite of what we would call hardships according to 2022 standards, Edith and Ralph established a Christian home in which three children were born, two of whom are the authors of this book, Clarence born in 1935, Melvin (now deceased) born in 1938 and Mardelle born in 1941. Ours was a home with no angry outbursts, no violence, no hitting except an occasional paddling, no drinking and no smoking. We are sure there were disagreements but only once did it become evident to me, Clarence. There were no angry words between my Mom and Dad but the silence was deafening.

While I, Clarence, enjoyed Sunday school and church, I dreaded seeing Sunday arrive because I was required to wear my suit make of reprocessed wool which made me itch for an entire two hour or more period until I could get it off and get into my overalls. Winter time was a little more tolerable because I could wear long underwear underneath so the wool suit was not touching my skin. I was in my glory when in my junior or senior year of high school I was able to convince my parents to buy me a blue surge suit that was made of a smooth fabric that did not itch and for the astronomical price of forty dollars.

During the research for this book we discovered that Ralph had a new barn and silo built in 1937. I, Clarence, had always thought that the barn and silo were there when Dad bought the farm, but then I was only two and a half when the barn was built so I guess I wouldn't remember.

By about 1940 or 1941, Ralph and Edith had made the decision to build "the new house." Lars Henning was the carpenter who would build the new house. These are my, Clarence's, recollection of building the house. My co-author, Mardelle was not born yet. Lars was a Dane and he and our Dad had become good friends. I remember that Lars and Nora would come for visits occasionally, maybe related to planning on the new house. These were unsettled times because the United States was becoming involved in World War II. I remember that there was concern that we might not be able to get

New barn and silo, left.

Below, the farmstead, sometime between 1937 and 1941.

the materials we needed to build the new house. The whole country was becoming geared toward supporting the war effort. I remember a sense of urgency in getting materials purchased before the supply was shut off. Lars had staked the location of the new house and they were ready to dig the basement.

The basement for the new house was dug with a crane on May 21, 1941. While that provided what was needed for pouring foundations and laying block, it left piles of dirt all around the site. While I had originally thought that Dad had dug the whole basement with a team of horses and a slip scraper, pictures that we have found prove me wrong. But I distinctly remember Dad using a team of horses to move dirt, using this piece of equipment that looked like a large shovel (3 or 4 feet square) with handles on one side like a wheelbarrow. With each pass the shovel would fill with dirt and when they reached the location where Dad wanted the dirt he would let go of the handles, it would turn upside down, and dump the dirt. I guess my six year old brain partially failed me.

I don't remember much about the construction of the house, except that my Dad worked along with the carpenters as much as he could and still get the farm work done. The cedar shingles were all individually dipped in a green colored stain and allowed to dry before they were used. The interior walls were of lathe and plaster. The plasterer's name was Jack Malnory. I remember that because of a rhyme we would say:

I'll tell you a story
about Jack Malnory
and now my stories begun,
I'll tell you another
about his brother
and now my story is done.

I remember a concern for exactness in laying the brick. The main color of the brick was strawberry but every few bricks,

Digging the basement.

a brick with a yellowish color was inserted. I remember my Dad watching that carefully and helping to line up the bricks in the proper sequence, especially on the front side of the house.

Moving day was an exciting day, and although the move was only a distance of twenty feet or so, I got up early, put all my toys in a box and moved them to the new house. Then off to school for me. When I came home in the afternoon, the move was complete, and we lived in a new house.

I have no recollection of the old house getting torn down but somehow it was gone and in its place a two-car garage was built. What I do remember is that there were honey bees living in the wall of the old house. Someone,

The old house & the new house.

and if memory serves me right, I think it was Dr. Wirtz, our dentist who made arrangements to come to the farm on a weekend, they would remove enough siding to find the queen bee and remove her to a hive, assuming the worker bees would all follow. Dr Wirtz, I believe, considered himself an amateur beekeeper and came with all the needed equipment. Well, they did not find the queen bee but they found washtub after washtub of honey. Obviously we had honey to eat and honey to share and bees wax to chew.

Clarence Remembers Growing Up on the Farm

Life on the farm was exciting and carefree for a young boy on our Iowa farm. We spent most of our time outdoors, summer or winter, romping and playing and helping, or so I thought, without the constraints on children so necessary in our 2022 world.

There were milk cows to take care of. In the beginning we had ten or so milk cows that we would milk by hand morning and evening, and when I was old enough I would help with the milking. Sometimes that was a challenge to avoid being slapped on the head by a wet cow's tail or being kicked by a touchy heifer who was getting used to the routine, all while sitting on a one-legged milk stool with the pail of milk between your legs. The milk would then be separated into cream and skim milk since only the cream was marketable for the production of butter and cheese.

I have no recollection of how the cream was kept cool in between trips to the creamery in Latimer; perhaps in a cold water tank. There was no market for the milk but it made great food for the hogs and chickens. It was Mother's job, I believe, to wash the many parts of the cream separator in preparation for the next milking.

As time went on, the cowbarn was remodeled so that the milking stanchion area had a cement floor with a gutter for the manure and a litter carrier to make the barn easier to keep clean. That was a great help in the winter time when the milk cows spent most of their time in their stanchions and only went out long enough for

Carefree farm children — Clarence, Melvin & Mardelle

them to get a long cold drink of water and for us to get the barn cleaned and fresh straw strewn where needed for bedding.

Remodeling of the barn was great fun for me because once the old floor was removed it was my job along with Maude, the draft horse that I would ride bareback around the farm, to ride round and round on the loose soil to pack it down in preparation for pouring cement. Up

Milk cows.

and down, round and round I rode until the dirt was compacted by Maude's large hooves and her weight. As I remember, the remodeling allowed for milking of a few more cows and made the whole job easier. At some time later a milking machine system was installed so no more milking by hand.

The other half of the barn was a pen where the young growing cattle born the previous year would be housed in the winter time until they could be turned out on pasture in the spring. And on the end of the barn were the stalls for the draft horses.

The haymow was equipped with a track on which the grapple fork apparatus would travel. The whole operation was powered by a rope attached to that apparatus and the grapple fork that ran through a series of pulleys to

Ralph's horsepower.

the south west corner of the barn. That rope would be attached to a team of horses and when in early years the loose hay and later years hay bales, were secured by the grapple fork, the team would pull the hay up to the hay mow where it was dropped in place. On occasion, twice that I remember, that rope would break. Dad was then off to Hampton where he would pick up a man he knew who was expert at reweaving rope. He would take off his shoes, climb into the hay mow and sit in the hay while he reweaved the rope. We were then ready to make hay again.

For at least the first ten years, if my memory serves me well, all the farm work was done with horses. There was an old Case tractor with lugged wheels, that may have come along with the farm when it was purchased, that sat unused except for running a belt to the saw for sawing wood, but the field work was done with horse power. A tragic mistake, or perhaps just fate, changed the Ralph Fitz farm from horse power to tractor power.

It was common practice sometime in the spring, for the cattle and horses to be turned out after being confined most of the winter, to romp in the fields that had been harvested the fall before. On this particular occasion both cattle and horses were turned out into a stubble field where corn had been harvested the previous fall. Within a very short time all of the horses were sick and dying. The veterinarian diagnosed it as

The Case tractor.

moldy corn poisoning. There wasn't much that could be done; the damage had been done.

Only Maude, the youngest of the horses was saved. She was still light enough that with the help of other men and a block and tackle, she could be raised up to her feet for a while each day until she finally could stand on her own. I can't remember any details or perhaps never knew any details, but before the planting season started, a shiny new John Deere tractor took over the job of field work.

In addition, there were the hogs and sheep to feed, an occasional orphan lamb to bottle feed, young calves to feed and chickens to feed. It was Mother's job to gather the eggs and make sure they were clean and ready to take to market. I have no idea why one summer we would have two geese. I wasn't fond of these geese because they would chase us kids and try to peck us. In the spring we would get several hundred baby chicks that would have their own building to grow up in. They took special care, needing supplemental heat until they feathered out. I remember one particular occasion when in the morning it was discovered that an animal had gotten into the brooder house and killed several of the young chicks. Was it a mink, or a skunk? We never knew. Ralph would buy mostly female chicks to replace the aging egg producing hens in the chicken house, but he always made sure there were enough roosters included to provide for those delicious Sunday fried chicken dinners that Edith, our mother, was famous for.

It was Ralph's job to chop the roosters head off using a machete. Once the roosters head was by chance in the fateful position on the stump, with a swift chop the rooster was left to jump around the yard until it bled to death. It was then time for mother to take over, first dunking the chicken in boiling hot water to loosen the feathers, and once the feathers were plucked, to singe the body hairs off the rooster with a flaming piece of newspaper. Once the entrails were removed and the chicken was washed, there was always one final look to make sure no pin feathers had been missed before the chicken was cut into pieces and made ready for the frying pan.

Ralph and Edith believed in education and even though Ralph had only achieved a tenth-grade education and Edith only an eighth-grade education, they had been good conscientious students and they made sure all three of us children got a good education. Mine, Clarence, started in a one room kindergarten to eighth grade country school that was located on the southeast corner of the section to the east of our farm, so I could walk to school either by going through our own field or walking the half mile on the road. Miss Nelson was our teacher and Josie Nelson and Marlo Nelson were my classmates, at least in the beginning. But for whatever reason Josie was moved up a grade so that left only Marlo and I in our grade, which was great because now my whole class could sit in one double desk.

Schoolhouse where Ralph started school in Greene Co.

Lunch was whatever we brought from home. At 11:30 Miss Nelson would announce that anyone who brought hot lunch could now set it on the heating stove to warm up. Bathroom was an outhouse where one always needed to make sure the person who used it previously hadn't set a booby trap like a can a water

[COPYRIGHTED 1900.]

TO THE PARENT OR GUARDIAN:

This report will be sent for your inspection and signature at the end of each school month. Please read it over carefully and TAKE NOTICE OF the following points:
(1). ATTENDANCE—Days of school and days absent.
(2). PUNCTUALITY—How often tardy during the month.
(3). DEPORTMENT—If low, learn why.
(4). BRANCHES PURSUED—Which ones.
(5). STANDING—In the various branches.
(6). Whether the pupil is ADVANCED ALIKE IN ALL HIS STUDIES. If he is behind in any branch, cooperate with the teacher in having extra work done in that branch till he catches up.

Please encourage the pupil:
(1). To attend EVERY day of school.
(2). To be on time ALWAYS.
(3). To do ALL the regular work of the course and thereby earn a diploma.
(4). To respect the AUTHORITY of the school.

If COMPLAINTS reach you, see the teacher at once. Give the teacher credit for being sincerely interested in the welfare of your child and don't wait for complaints, but keep in touch with the school by making FREQUENT VISITS.

THE PUPIL should keep this report as a precious and valuable souvenir of this year's work. Show it to the next teacher.

CAUTION TO THE TEACHER:—Do NOT WRONG THE PUPIL BY MARKING HIM ABOVE HIS ACTUAL MERITS. Let your tests be fair and sufficient. Honest patrons will sustain the honest teacher. Fill out remark column.

The parent will please sign below and return promptly to the teacher. SUPERINTENDENT OF SCHOOLS.

MONTH
1 Mrs. Alda E. Fitz
2
3
4
5
6
7
8
9
10

TEACHER'S

Report to Parents

SCHOOL *Logan, No. 2*
Calhoun County, *Iowa*
Pupil *Ralph Fitz*
Age *11* Grade or Year of Course
For School Year Ending _____ 190__

Teachers to whom pupil recites during the year:
Amy B. Eshbaugh

Superintendent. (Teacher insert name.)
N. R. Sandy

Form 8.
Published by
HAMMOND BROS. & STEPHENS,
FREMONT, NEBRASKA.
PRICE, ONE CENT.

Report for the Year beginning *Sept. 4* 190*2*, and ending _____ 190__

BRANCHES PURSUED, ATTENDANCE, DEPORTMENT, ETC.	YEAR OF COURSE OR STUDY ENTERED	MONTHS WORK TAKEN UP	Sept 29	Oct 27		SUMMARY FOR						SUMMARY FOR	YEAR OF COURSE FINISHED	MONTH FINISHED	REMARKS ON WORK DONE AND RECOMMENDATIONS OR PROMOTIONS FOR NEXT YEAR, ETC.
Orthography			89	87											
Reading			90	90											
Writing			92	93											
Number															
Arithmetic			89	85											
Language															
Grammar			86	89											
Geography			87	85											
Physiology															
History															
Civil Gov't															
Days of school			20	20											
Days absent			2	1											
Times tardy			0	0											
Deportment			98	99											

Ralph's report card.

balanced on the door so it would dump on the next person through the door. Recess was unorganized yard games like "Beckon Beckon, who's got the Beckon," an adaptation of hide and seek, and when the school bell rang recess was over. We would study or do projects at our desk but recitation was done in the front of the room under the periodic table and the picture of George Washington that adorned the wall.

We had a pretend store stocked with empty boxes and cans where we would learn about merchandising and how to make change, a lost art to our current generation. Education was at that time based on reading, writing and arithmetic rather than indoctrination which seems to dominate 2022 education. Hours were spent on phonics and sounding out words as well as penmanship.

I guess I was paying attention during the phonics lessons because I qualified for the county spelling bee only to be eliminated by misspelling the word "restaurant." "Restaurant" was not a familiar word to me since we seldom if ever went to a restaurant as a family. With homegrown beef and pork, a large vegetable garden, a pantry full of goods mother had canned, a root cellar with potatoes, carrots and parsnips and corn for popping in addition to mother's fried chicken there was no need for a restaurant.

One of my fondest memories of country school was VE Day, May 7, 1945, when early in the morning Hitler and Nazi German had surrendered which brought war in Europe to an end. By the time I got to school that morning the school bell was already ringing, and taking turns, we kept it ringing until the last person left at the end of the school day.

Between third and fourth grade the country school was closed and I was now picked up by the Rowan school bus. After two years in the Rowan school, consolidation put me in the Franklin Consolidated district in Latimer. But I had made good friends in the Rowan school and I don't know what arrangements were made, but I was allowed to walk a half mile to catch the Rowan bus on the consolidation line to complete the sixth grade in Rowan. But it was short-lived and I was off to make new friends in seventh grade at Franklin Consolidated where I finished my middle and high school education.

We hadn't been in the new house very long when on a Sunday morning, December 7, 1941, news came on the radio that the Japanese had bombed Pearl Harbor. I, Clarence, was only six years old so I didn't hear that report. Grandma and Grandpa Hill were coming for Sunday dinner after church, the traditional Sunday fried chicken dinner. In those days, at least on Iowa farms, dinner was at noon, supper was in the evening and lunch was that quick sandwich and a drink in the middle of the forenoon and again in the middle of the afternoon.

During heavy field work, the ladies would bring the lunch to the field so the men could get a quick break without wasting time coming out of the field. I fondly remember how good that bologna sandwich tasted washed down with ice cold Kool-Aid. But on this Sunday morning there was tension in the air. I distinctly remember being in the kitchen of the new house with Mother and Grandma.

It seems like I was about the same height as the kitchen counters because I had to look up to see the fear in their eyes. I didn't really know what had happened but I knew it was something bad. Little did I know that this act on the part of the Japanese would bring the United States into the world war and would impact greatly life on our farm.

I doubt that Ralph ever knew, or if he did I never heard him mention it, that with the massive damage and death inflicted by the attack, his ship, the USS *Allen* had been present at Pearl Harbor on the day Japan attacked but escaped any damage.

On Dec 8, 1941 the United States declared war on Japan. About three days later Germany and Italy declared war on the United States. The United States was now a full participant in World War II, which would have major effects on the country and on Ralph Fitz's farm. In most ways our farm was not negatively affected by the war, the most personal being that we had no family members who were of military age, but many were not so lucky and 330,000 men would never return to their families.

Farming was an important industry in support of the war effort because farmers were the food producers of the country including our military personnel as well as producers of needed nonfood commodities. With most of the young men away at war there was a real shortage of farm workers. I remember that someone, and I think it was Lars Henning the carpenter who built our new house, that asked Dad if he could help him plant corn as his contribution to the war effort. Because he knew Dad was short-handed, Lars planted the field just west of the farmstead.

Dr. Wirtz, our dentist was away at war so our dental care suffered until the war was over. Rationing was a hardship but not severe. We had plenty of food because we grew our own food.

As a youngster the thing that impacted me the most was sugar rationing and my reprocessed wool Sunday suit that I mentioned earlier. My dislike of the taste of artificial sweeteners started with trying to sweeten cereal with saccharin, which tasted awful. I also missed chewing gum, which all went to the soldiers and sailors.

I remember that in fourth or fifth grade it was traditional for those who had birthdays during the school year to bring treats for the rest of the class, which was probably about twenty. I remember that Mom and I had to go to several stores to get enough candy bars, because there was a limit on how many they could sell to any one customer.

Even us school kids got involved in the war effort. I remember walking the road ditches in search of milkweed pods which we were told were used to make life preservers. Perhaps the butterflies were the losers. I also remember taking apart an old treadle sewing machine that was laying in the grove and by donating fifty pounds of scrap metal I got to ride around the city square in an Army jeep. Big stuff for a young boy.

One growing season Dad grew hemp to aid the war effort. It was to be used for making rope, we were told, but it was also a good cash crop for the farmer who grew it. It was nasty stuff to harvest but it got done.

Dad usually planted a field of sweet corn, I believe through a contract with the Marshall Canning Company in Hampton. Here again, it was intense at harvest time because it needed to get to the canning company quickly whenever the canning company inspector said it was at prime flavor, but was a good cash crop. On at least two instances during the war I remember having help in picking sweet corn, which was all done by hand, by workers brought in by the canning company.

One such worker was Tinker Alexander, a migrant worker from Jamaica. Tinker lived with us until the sweet-corn harvest was done, and was our first up close exposure to a non-white person. Another year the canning company brought in German prisoners of war to help with the harvest. The prisoners wore shirts with a large POW printed on the back and Ralph had to pick him up at the canning factory each morning and return him after the day's work was done so he could be kept under armed guard until the next morning.

Farming was dramatically different during my childhood than it is now. Farmers were pretty much on their own to get their crops planted and tended up to harvest time, but harvest brought a good deal of working together, especially in regard to threshing and silo filling.

While oats was not a great cash crop, it made a good nurse crop for alfalfa or clover as those plants were becoming established. When the oats was mature a binder was used to cut the oats and bind it into bundles. Then it was time to shock the bundles and as soon as I was able I could help with that. Each shock had a certain number of bundles leaned against one another with one bundle on the top to form a cap. The idea being to keep the oat grains, as much as possible out of the weather and moisture which would damage and discolor the grain. When the oats was all shocked, it was time for the threshing run.

One farmer would usually own the threshing machine, sometimes called a separator, because it would separate the oats from the straw by shaking it and then elevating the grain into a wagon and blowing the straw onto a stack. The threshing machine was run by a tractor that would run the belt that turned the threshing machine.

My first memory of threshing at our farm was a steam engine serving the purpose of running the threshing machine. But first the bundles had to be hauled from the field to the machine and that is where I came in when I was old enough. I was in my glory when I could work as a spike pitcher going from wagon to wagon in the field and pitching bundles onto the horse drawn hay rack with a pitch fork.

As the team of horses pulled each wagon close enough to the threshing machine so the bundles could be pitched into the machine there was sometimes a bit of excitement if the team of horses became spooked at getting this close to the noisy threshing machine. As times moved on the threshing machine became a part of the discarded technology destined for antique status and combines came into use which could do the whole job in the field.

During the threshing run the farm wife was expected to serve dinner (remember that dinner was at noon) to the threshing crew. Threshing was a dirty job and provisions were made to accomplish that without all of the dirt entering the house. I remember at least one occasion where a washing station was set up in the yard with water and soap for washing and towels for drying before either entering the house of eating picnic style in the yard. A fun remembrance is that when the threshing crew was at our neighbor, George Park's farm, we could always count on having tomatoes and peaches for dinner. George's wife, Dorothy, revealed at some point that that was the only vegetable and fruit that George would eat.

It was at George's place that I first tasted beer. The beer was in a tank of water to keep it cool and a neighbor boy and I decided we would try it while the men were all busy elsewhere. It had an awful taste so we decided to mix it half and half with water, but that didn't help so I think we decided beer was not for us.

The other operation that was usually done with a crew was silo filling. One farmer usually owned the silo filler and would go from farm to farm along with the crew. The filler was a machine that would chop the corn bundles and blow the chopped corn to the top of the silo where it would fall into the silo until it was full. At the proper stage of development, when there were ears of corn developed but the stalks were not too dry, the corn stalks would be bound into bundles by a machine called a corn binder. These bundles would be hauled to the silo and fed into the silo filler.

As the chopped corn stalks were compacted and with time, a fermentation process took place, not too much unlike sauerkraut and the result was a sweet-smelling silage that the cows seemed to relish. My only involvement with silage, when I was old enough, was to climb to the top of the silo through an enclosed ladder and pitch enough silage down for that feeding. We had no idea that we were breathing a toxic gas. In 2022 I believe most silage is stored in large plastic bags laid on the ground.

Growing corn was pretty much done by the farmer and sometimes a hired hand. Corn was checked during my childhood which meant that the corn was planted in hills rather than being drilled as is common in 2022. It was done with a team of horses pulling a corn planter. For each row, or more depending on the planter, a wire with knots in it every so many inches was stretched across the field. As the planter reached each of these knots, the planter would drop usually 3 kernels of seed corn into the furrow that was dug in the front of the planter and covered in the rear of the planter.

The weeds were controlled by going down each row with a cultivator drawn by a team of horses. I remember one incident that was short lived.

I don't know how old I was but while Dad was away on an errand, I had harnessed a team of horses, hitched them to the field cultivator and was cultivating a cornfield. When Dad returned and discovered what I was doing, it became obvious that Dad did not want that field of corn cultivated that day as he took control of the team and I was sent home. Sometimes, I thought I was more grown up than I actually was.

Harvest was most likely done by hand in the beginning, but my first remembrance was picking corn with a tractor drawn corn picker that would pick the ears and elevate them into a wagon attached to the rear of the picker. There are gaps in my memory involving corn harvest probably because I would have been in school at corn picking time. I do remember unloading wagon loads of corn and elevating it into the corn bins, the elevator being run at first by a team of horses being driven round and round to operate the elevator and later by a tractor with a belt attached to the elevator apparatus.

The war effort had a big influence in bringing the country out of the economic depression that it was still not recovered from at the beginning of World War II, and farming was well positioned to be a benefactor. Food and fiber were needed commodities for the war effort and that's what farmers

Elevating corn.

produced. While we do not have actual data to support it, it seems that Ralph and Edith's farming operation, after having struggled during the Depression years, was doing pretty well.

When Ralph took final possession of the farm in 1934, we believe the farmstead was comprised of the house, the chicken house, hog house, corn crib and several small buildings. Within a few years the barn and silo were built, then the new house, and sometime after that, a tenant house for hired help, a cattle shed and a machine shed as well as numerous small structures for specific purposes. The building spot was being well cared for.

I, Clarence, remember an amusing incident. Ralph, my Dad, had hired a painter to paint the chicken house. And I, always being inquisitive, would wander over to where the painting was happening and the painter would stop painting and talk to me. Dad must have noticed because I was told not to go near the painter because when he was talking to me he wasn't painting.

Ralph was obviously a good manager with so many profit centers going on at the same time and always being at the mercy of the weather as all farmers are. In addition to producing the commodities he had to analyze the various markets and sell at the best time to realize profits. He had learned to be frugal during the depression years which was to his benefit if all did not go as he had hoped.

Mardelle Remembers Growing Up on the Farm

It was October 8, 1941, the Henning construction crew was at the farm working and ready for dinner, but they had to wait because I wouldn't. I, Mardelle, was born at 12:45 p.m. Now Daddy had his two sons to help on the farm and his little girl to spoil. Mama & Daddy were good friends of Lars & Nora Henning and would visit often. Lars would always remind me that he had to wait for dinner until I was born.

When I was old enough to help, I would help Mama with her chicken chores, gathering eggs, which I was never happy when the hens were on the nest and would peck me or fly off the nest. After carrying the pail of eggs to the house in the basement I would help wash the eggs, one by one, to be ready to be put in an egg case to take to market.

I would help Mama clean the inside of the chickens (I'm glad Mom can't see how they are cleaned nowadays when bought at the grocery store) and help cut the chickens in serving pieces ready to fry. Mama cooked the best fried chicken and still is remembered by cousins at family reunions. I also remembered Mama's chickens always had more than two legs in that frying pan!

Washing clothes and cleaning the house were other chores I helped Mama with. We had a wringer washing machine and hung the clothes on the clothesline, summer and winter. I remember in the winter when the clothes were dry they would be froze solid, but they smelled so good. Then there was ironing to be done to all.

Mama loved her popcorn. White & yellow popcorn seeds were planted every year. After being picked and dried I remember helping her shell the corn on a hand corn sheller in the machine shed. Mama was very particular with her corn; she would check the dryness by popping a little ever so often to make sure it popped the way she wanted…perfect! She then stored it in large glass jars.

In the summer Daddy would let me bottle feed a couple of orphaned lambs. When they were taken to market, I got the money that they brought.

After becoming a teenager Daddy let me drive the John Deere tractor on the baler when baling hay. That was one of my favorite jobs to help Daddy.

Feeding my lamb.

Yes, I did help stack the bales on the rack once in a while, but driving the tractor was more fun. I did help milk the cows by hand, but that was short-lived.

We had a big garden on the farm and I would help with hoeing weeds, picking strawberries, raspberries, green beans and peas. I think it was the men that did the digging of carrots, potatoes, etc. Although not edible Mom would have several rows of gladiola flowers that were beautiful in the garden closest to the road so everyone could enjoy those gladiolas.

I, Mardelle, was very lucky to be born in a brand new house with all the conveniences of a modern home.

In 1965 Ralph and Edith built a new house in Latimer, Iowa, had a farm sale at which they sold the dairy herd, and at age 72 Ralph quit farming and moved to town.

CHAPTER 10

THE WOMAN HE CHOSE TO SHARE HIS LIFE

We will start this chapter with a letter received by my daughter Karna, Clarence's daughter, from her grandma as part of a school project in 1977.

Dear Karna,

I, Edith Aleta Hill Fitz, was born on a farm four and one half miles north of Dows at the home of my parents Clarence and Mildred Bates Hill on August 28, 1903. The mothers didn't go to the hospital to have their babies those days, so I was born at home in Wright County, Iowa. A Doctor George from Dows came to my parents home. He had to drive a horse and buggy as hardly anyone had cars then. I guess I growed up pretty much like babies do today. When I was ten months old I got whooping cough and my parents said I almost died but guess I was tough because I didn't. When I was five years old I started to school in a one room country school one and a half miles from our home. We usually walked to and from school. We didn't have school buses like you do now so we had to walk. When the weather was bad our father would hitch a team of horses to a wagon or buggy and take us to school or get us at night. We liked it better when he used the buggy as the wagon was so bumpy and you usually had to stand up in it. When school was out in the spring we would always have a picnic and all the parents would come and bring food. And then after all the schools in the district were out we would have a big picnic with all the schools. We always had it on the Ellsworth Ranch which was four or five miles from our home.

When I was nine or ten one of our neighbors got a car. When they drove past our school all of us kids would stand and watch it until we couldn't see it any longer. We couldn't imagine a horseless carriage and everyone was almost afraid of it.

I can remember going to church in the school houses when I was pretty young, then of course when we were older we always went to the Methodist Church in Rowan, and that is where I still go.

Several of the neighbors around our home would haul their milk or cream to Dows and it was always a big thing if one or two of us would get to go to Dows with our father when he hauled the milk.

The year that I was twelve my mother and three of my aunts had a surprise birthday party for me. It was a nice day and so many girls and boys came and we played party games in the front yard. That was a big day for me. Then that night my father and uncle had to go to Dows to get a lady who was coming on the train so my cousin and I asked if we could go along and we got to go. We went in the restaurant and each had a dish of ice cream. That's the first time I ever had ice cream in town so I still remember it.

Also that same year a man came to our home to try to sell my father a car. He wanted to give us a ride in it. He said they were having a circus at Iowa Falls and wanted to take us to that. So my Mother, Father, brother Lawrence and I got to go to the circus. I don't think I will ever forget that as I had never been to a circus before and that was a big one. But my father did not buy the car.

I had eight brothers and sisters younger than me and one brother older than me. They were:

Lawrence Edward Hill born October 30, 1901
Edith Aleta Hill Fitz born August 28, 1903
Laurel James Hill born May 29, 1905
Veva Jessie Hill White born May 5, 1907
Clair Cecil Hill born October 24, 1909
Mildred Marcella Hill Pepper born July 15, 1911
LaVonne Esther Hill Morris born March 30, 1913
Arthur Truman Hill born December 2, 1915
Evelyn Ida Hill Boyington born April 28, 1917
Aubrey Arlie Hill born January 14, 1920

My youngest brother Aubrey died in 1968 and my oldest brother Lawrence died in 1971.

On March the 14th, 1934 I married your grandpa Ralph Glenn Fitz and moved with him to the farm on highway three where your Uncle Melvin lives now. We had three children born to us. Your daddy Clarence Ralph Fitz born February 18, 1935, your Uncle Melvin Glenn Fitz born February 2, 1938 and your Aunt Mardelle Marie Fitz Meyer born October 8, 1941.

We got electric lights in our home when Melvin was about 1 ½ years old. We still had to carry all the water we used from the well and heated it on the stove, which was a stove that you burned wood, cobs, or coal in. We would pick up cobs and wood in the hog yards and in the grove to burn. We didn't have bath tubs like you do now days. We would bring a big round tub in the kitchen so it would be warm. Then after we warmed the water on the stove we would pour it in this tub and take our bath.

In 1941 we built the house on the farm that Melvin's family lives in now. Then we had running water, electric lights, bathroom and all. It didn't seem possible to have all those things but we enjoyed them. We still didn't have a refrigerator or electric cook stove for a few years after that though, but then we got a deep freeze and used that for a year or so, then got the stove and refrigerator. So then we were pretty modern.

My father Clarence Hill died September 5, 1952 at the home where I was born and my mother then moved to Dows and in 1966 she had a bad stroke and she died in the hospital at Clarion on October 24, 1966.

Your Grandpa Fitz and I built a new home in 1965 and 1966 and moved into it the last of May 1966. We had lived on the farm where Melvin lives from the time we were married until we moved to Latimer. We lived together here in Latimer eleven years. We worked hard but we had a lot of fun too and a real good life. On May 4, 1977 your Grandpa Ralph Fitz died. So now I'm trying to learn to live alone and it isn't easy, it's so lonely.

Love, Grandma

Clarence Hill, your author's grandpa on our mother's side, was born July 1, 1865 at Elkader in Clayton County, Iowa. He was the oldest son of six children born to Elijah and Adelia Hill. His mother died when he was seven years old and at age fourteen he was taken into the home of Grandma McCallum.

Clarence married Mildred Bates on February 6, 1901. Mildred's parents were James Bates of Popejoy and Marcella Dent.

Edith, being the oldest daughter, had the responsibility, perhaps self-imposed, of helping her mother. That was never more true than on Christmas Day. Santa had come, and there were gifts under the tree to be opened, but right after the morning farm chores were done, we were off to Grandma and Grandpa's house so mother could help grandma get the turkey roasted.

By the time the dishes were done, and there were a lot of dishes with ten children, their spouses and all the kids, it was late afternoon. Remember, that was before the advent of paper plates and plastic silverware. And then, of course, there were cows to milk and the other chores to be done. Torture for an impatient child, but finally we could have our family Christmas and find out what Santa had brought the night before.

Grandma and Grandpa's Christmas tree deserves remembering. It was obviously just a branch cut off a pine tree and decorated. Sort of like Charlie Browns Christmas tree. One of the decorations I, Clarence, remember was a small candle holder with a pinch clamp to be attached to a branch. That was probably the forerunner to Christmas tree lights but terribly dangerous if anyone ever lit the candles.

One such time when Edith stepped up was when she discovered that Grandma, due to age, had not taken her hair down from the sort of bun that she always wore for who knows how long and it had become terribly snarled. Edith spent several afternoons tediously working on the snarls until finally they could run a comb through Grandma's hair.

Our family, Ralph and Edith and kids, would often go to Grandma and Grandpa Hill's house for a visit on Sunday afternoon. And I, Clarence, would invariably decide it was time for me to inventory all the animals on Grandpa and Grandma's farm. On one occasion I brought my list in to show it off, and I vividly remember Grandpa Hill sitting on the couch and looking at my list and saying, "This boy will go to Ames." "Ames" was reference to Iowa State College located in Ames, Iowa, and that planted the seed in my mind for attendance at Iowa State College and for my eventual graduation from what by then had become Iowa State University.

Edith quilting.

Above all else, Edith was kind. Perhaps reticent, but kind. In some ways she was shy, but more often it was deferring to her husband whom she respected to a fault. She was a Christian lady who was faithful to her church and to the Ladies Aid of which she was a part. She was an accomplished cook and baker and was in her glory when she could lay out a spread before friends or family. And when there was time she would sit down at the piano and play.

She was not an accomplished pianist but she could read music and when it came to the hymn "Star of the East" she was flawless.

Edith was an accomplished seamstress and produced some beautiful crocheted doilies, dresser scarves, knitted some gloves and stockings, made quilts, all in the evening after the days work was done. On one occasion, after Ralph had died, she was asked by the local school to give a demonstration to the home economics class on tatting. Tatting, a lace making procedure, is pretty much a lost art but someone found out that Edith knew how to do it.

Edith loved her house, her home, and was very comfortable just staying there. But Ralph liked to travel and see new things and new people, and of course she would go with him to places even as far away as Israel or as primitive as Oak Island in Lake of the Woods.

This was our mother, the authors of this book, and the woman that Ralph chose to share his life.

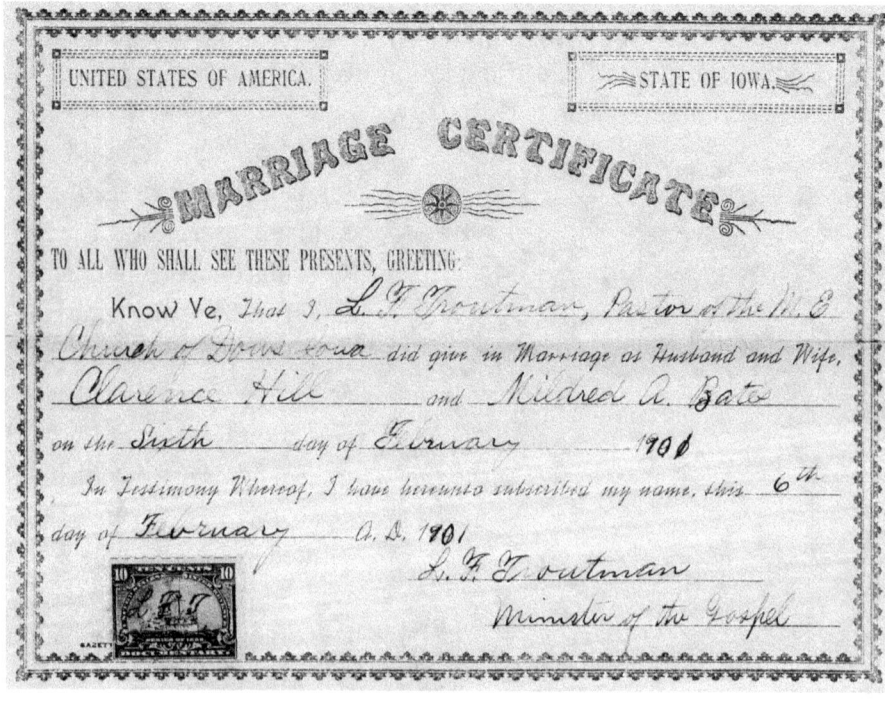

Clarence and Mildred Bates' marriage certificate.

Wedding picture of Clarence Hill & Mildred Bates.

*Grandpa Hill, center, Grandma Hill on the left, and Grandpa Hill's sister
Clara on the right.*

Mary Hill Vansickle,
Grandpa Hill's half
sister.

Uncle Lawrence and our mother.

Grandma Mildred, back row center.

Lawrence, Edith and Laurel (mother at 2+ plus years old)

The Hill family. Back row: Lawrence, Laurel, Clair. Center row: Arthur, Evelyn, Edith, Veva, Mildred. Front row: Grandpa Hill, LaVonne, Aubrey, Grandma Hill.

Grandma and Grandpa Hill's 50th wedding anniversary.

Edith.

Mildred Hill
Dies at Age 87;
Lifelong Dows Resident

Mildred A. Hill, 87, longtime resident of the Dows area died at the Clarion Hospital Monday, October 24th after a lingering illness.

Mrs. Hill was born August 16, 1879 in Marshall County, Iowa, the daughter of James and Marcella (Dent)Bates. She came to the Dows area to live in her youth. She was married to Clarence Hill, February 6, 1901 at her parents home southeast of Dows. They purchased a farm home northwest of Dows where they spent the rest of their married life. Mr. Hill passed away in 1952. She remained on the farm until 1954 at which time she moved to Dows. Mr. and Mrs. Hill celebrated their Golden Wedding anniversary in 1951. They retired from active farming in 1939 but remained on the farm.

She is survived by ten children: Lawrence of Dows; Edith, Mrs. Ralph Fitz of Latimer; Laurel of Dows; Veva, Mrs. Clifford White of Dows; Clair of Rowan; Mildred, Mrs. Vernard Pepper of Independence; LaVonne Mrs. Dewain Morris of Ellendale, Minn.; Arthur of Iowa Falls; Evelyn, Mrs. Wilson Boyington of State Center; and Aubrey Hill of Ellendale, Minn.

The following brother and sisters also survive: Elmer Bates of Dows; Mrs. Jessie Green of Traer; Mrs. Theda Fanselow of Dows and Mrs. Cecil Buttenoff of Topeka, Kansas. Thirty-eight grandchildren and thirty-seven great grandchildren also survive

She was preceded in death by her husband, her parents, three brothers and one sister and one grandson, Delbert White.

Funeral services were held Wednesday, Oct. 26, 1966 at the Rowan Methodist Church in Rowan. Burial was in Fairview Cemetery at Dows. Rev. Louis Aitken officiating. Van Hove Funeral Home was in charge of arrangements.

Grandma Mildred Hill's death notice.

Clarence Hill Dies near Dows

The last rites for Clarence Hill were held Monday afternoon in the Methodist church in Dows with Rev. C. Wilbur Egeland of the Rowan Methodist church officiating, assisted by the Rev. Donald Roberts of the Dows Methodist church. Interment was made in Fairview cemetery near Dows. Bearers were: Messrs. Earl Sellers, Karl Rietz, Marc Ihm, Jesse Peyton, Claud Johnson, W. G. Knox, Elmer Westlund and John Hawe.

Clarence Hill, the eldest son of six children, born to Elijah and Adelia Hill on July 1, 1865, in Clayton county, Elkader, passed away at his home northwest of Dows, Sept. 5 at the age of 77 years, two months, and five days.

His mother died when he was seven, leaving him without the comforts of a mother's loving care. At the age of 14 he was taken into the home of Grandma McCallum, as she was always called, where he remained until manhood. He was engaged in farming with his brother until the time of his marriage.

At an early age he was united with the Methodist church in Rowan and was a faithful member until his death.

He was united in marriage to Mildred Bates on Feb. 6, 1901, at the home of her parents southeast of Dows.

They purchased a farm west of Dows where they began their married life. After living there one year, they purchased the farm home northwest of Dows, where they resided at the time of his death.

Mr. and Mrs. Hill celebrated their golden wedding anniversary in 1951. He retired in 1939 but still remained in the farm home.

He is survived by his loving wife and children, Laurence, Laurel, Arthur, and Veva, Mrs. Clifford White of Dows; Edith, Mrs. Ralph Fitz of Alexander; Clair of Rowan; Aubrey and LaVonne, Mrs. DeWain Morris of Ellendale, Minn.; Mildred, Mrs. Vernard Pepper of Colo.; and Evelyn, Mrs. Wilson Boyington of McCallsburg; also 34 grandchildren. He was preceded in death by two brothers and three sisters and one grandson, Delbert White.

Those from out of town who attended the Clarence Hill funeral Monday were: LaVern Hill A/3c, Enid, Okla., Arthur Bates and Jessie, Mr. and Mrs. Virgil Eaton, Will Rice and Mrs. Daisy Poulson, Ruthven; Ray Bates, Benson, Ariz.; Mrs. Edd Green, Traer, Mr. and Mrs. Herman Wildeboer, Zearing, Mrs. Veva Wildt, Ft. Dodge, Mr. and Mrs. Fred Petersen, Story City, Mrs. Fred Miller and Mrs. Harmina Wildeboer, Hampton, Mr. and Mrs. Will Azeltine, Alden, Mr. and Mrs. Glenn Morris, Mr. and Mrs. Francis Morris and Mrs. Mabel Morris of Belmond.

Clarence Hill Pioneer Resident Laid to Rest

Funeral services for Clarence Hill, 87, pioneer resident of this community who died last Friday September 5, were held from the Methodist church here Monday with Rev. C. W. Egeland and Rev. Donald Roberts officiating. Favored selections were sung by Mrs. C. W. Egeland, accompanied by Max Banwell. Pall bearers were John Hawe, Earl Sellers, Glen Knox, Mark Ihm, Jess Peyton, Elmer Westlund, Carl Rietz and Claude Johnson. Interment was in Fairview cemetery with the Halvorson funeral home in charge.

Mr. Hill had sustained a severe heart attack about two weeks ago which made him bedfast and suffered several other attacks since that time until he succumbed on Friday, September 5.

Clarence Hill

Clarence Hill, the eldest son of six children born to Elijah and Adelia Hill on July 1, 1865 in Clayton county, Elkader, Iowa passed away at his home northwest of Dows, September 5, 1952 at the age of eighty-seven years two months and five days.

His mother died when he was seven, leaving him without the comforts of a mother's loving care. At the age of fourteen he was taken into the home of Grandma McCallum, as she was always called, where he remained until manhood. He was engaged in farming with his brother until the time of his marriage.

At an early age he was united with the Methodist church in Rowan and was a faithful member until his death.

He was united in marriage to Mildred Bates on February 6, 1901 at the home of her parents southeast of Dows. They purchased a farm west of Dows where they began their married life. After living there one year, they purchased the farm home northwest of Dows, where they resided at the time of his death.

Mr. and Mrs. Hill celebrated their golden wedding anniversary in 1951. He retired from active farming in 1939 but still remained in the farm home.

He is survived by a loving wife and children: Laurence, Laurel, Arthur, and Veva, Mrs. Clifford White of Dows; Edith, Mrs. Ralph Fitz of Alexander; Clair of Rowan; Aubrey and LaVonne, Mrs. DeWain Morris of Ellendale, Minn.; Mildred, Mrs. Vernard Pepper of Colo; and Evelyn, Mrs. Wilson Boyington of McCallsburg; also thirty-four grandchildren. He was preceded in death by two brothers and three sisters and one grandson, Delbert White.

He was always a kind and loving husband, father and grandfather and will long be remembered by all.

Grandpa Clarence Hill's death notices.

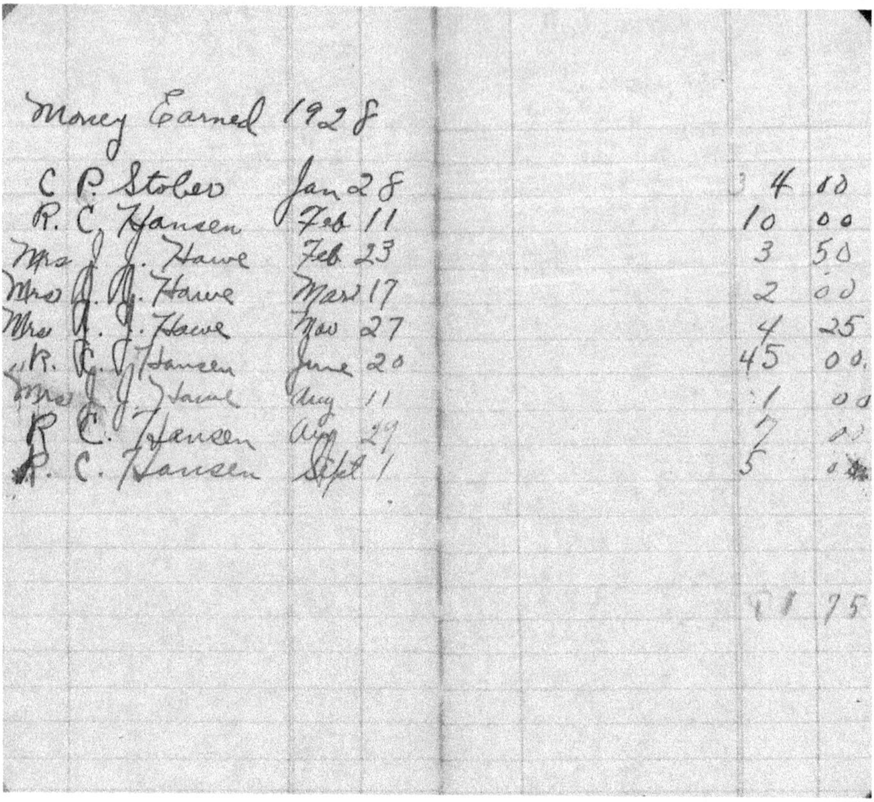

Money Earned 1928

C. P. Stoles	Jan 28	3 4	10
R. C. Hansen	Feb 11	10	00
Mrs. J. J. Hawe	Feb 23	3	50
Mrs. J. J. Hawe	Mar 17	2	00
Mrs. J. J. Hawe	Nov 27	4	25
R. C. Hansen	June 20	45	00
Mrs. J. J. Hawe	May 11	1	00
R. C. Hansen	Aug 29	7	00
R. C. Hansen	Sept 1	5	00

81 75

Edith's income for 1928.

January	1928			
Gum	3rd	Cash		5
M.S. Gundrum merchandise	4th	Cash	1	05
Richards & Rummel Book	4th	Cash		25
Variety store Candy	4th	Cash		15
B.D. Banps candy	15th	Cash		25
Pattersons gloves	16th	Cash		49
Grab & Run Candy	14th	Cash		10
Hougass co pillow tubing	21st	Cash		50
M.S. Gundrum merchandise	21st	Cash		50
Johnny Candy	21st	Cash		25
Post Office Stamps	24th	Cash		50
Hickley merchandise	26th	Cash		55
S.W. Whitten C. Thread	26th	cash		25
Jonesboro Gum	26th	Cash		10
			Total	
			$4	89

Edith's expenses, January 1928.

February 1928.

Barber Shop Clarion Haircut Feb 4	Cash		50
Variety store Clarion candy Feb 4	Cash		10
Shoe store Eagle Gr. rubbers Feb 10	Cash	1	00
Music store Eagle, Music Feb 10	Cash		35
Soda Fountain Milk Feb 10	Cash		5
Chicago mail order Goods Feb 13	Bank draft	4	58
Hearth & Home paper Feb 13	Bank draft		25
M.S. Anderson Merchandise 17	Cash		74
B.D. Buys candy Feb 17	Cash		15
Variety store dress pattern Feb 17	Cash		15
Petersons Shoe repair Feb 28	Cash	1	25
Variety Store merchandise Feb 28	Cash		25
Ticket to movie Feb 24	Cash		50
	Total	9	87

Edith's expenses, February 1928.

March   1928.

		Total 6.59	
Telephone Call 1004 Rowan	Mar 30	cash	.10
Drug Store Rowan Gum	Mar 2	Cash	10
Dr. E. E. Latta Filling	Mar 5	Cash	2 00
M. S. Gunderson Merchandise	Mar 5	Cash	60
Otis Johnson Merchandise	Mar 5	Cash	30
Fruit Store Hampton Fruit	Mar 8	Cash	55
A&P Hampton Cookies	Mar 8	Cash	19
Meat Market Onna Meat	Mar 9	Cash	10
M. S. Gunderson Thread	Mar 15	Cash	50
M. S. Gunderson Thread	Mar 15	Cash	25
Al Bailey Lettuce	Mar 17	Cash	20
M. S. Gunderson Gum	Mar 17	Cash	5
Russell Drug candy	Mar 17	Cash	10
M. S. Gunderson Thread	Mar 19	Cash	25
Centerparlor Marcell	Mar 22	cash	75
A&P Store candy	Mar 22	Cash	10
Drug Store Eskimo pie	Mar 22	Cash	15
O. D. Boys Gum	Mar 5	Cash	5
O. D. Boys candy	Mar 30	Cash	25

Edith's expenses, March 1928.

CHAPTER 11

DADDY'S GIRL REMEMBERS DADDY

Memories Of An 80-Year-Old Daddy's Girl

Mardelle Marie Fitz Meyer

I remember Mom telling me when I was little that Daddy took me with him in the truck to Latimer on business. He forgot to bring me home with him so back to town he went to get me at the store where he left me.

I was also told when I was little Daddy had this black man working for him and living with us. This man would put me on his lap and try to teach me the alphabet. When he came to the H he would say HELL.

I remember after Mama washed my hair she would wrap rags in my hair to make curls. When I was in second grade I got my first haircut and permanent. I remember hiding under my desk at school because I didn't want anyone to see me.

I remember Grandpa and Grandma Hill visiting and Grandpa asking my folks if they couldn't afford to buy me shoes (I never liked wearing shoes and still to this day I take my shoes off when coming in houses).

I also remember staying all night with Grandma and Grandpa Hill. After supper we would watch the Kate Smith TV show. Grandpa liked to listen to Kate Smith sing.

I remember Grandma and Grandpa Hill staying with us after they were injured in an auto accident. Grandma was on a hospital bed and she looked like she had a body cast on.

I remember going to Aunt Hazel and Uncle Ralph's in Waterloo to my cousin's weddings. They were beautiful and I remember Uncle Ralph crying at Jeannie's wedding, his last daughter to be married.

I remember Daddy putting a new chicken coop by our house for me to use as a playhouse.

I remember that at my brother's birthdays I, too, would get a gift.

My playhouse in the background.

I remember Mom having to make gravy for me so I would eat my potatoes. I think I mentioned that I was spoiled!

I remember taking piano lessons every Saturday in Hampton. When done I would walk to Aunt Myrtle's place and wait till Mom would pick me up.

I remember Mom playing the piano when she had time, but I don't think her and I ever played duets.

Speaking of Aunt Myrtle, every summer I would stay all night with her. She always fixed creamed tuna on toast for me. That was a treat for me because Mom never made that at home.

Aunt Myrtle Fitz

I remember having lots of cats on the farm, but Tommy was my favorite. He would get in a lot of fights at night so I would put bandages on him so he would heal. He was very patient with me when I would put doll clothes on him and push him around in my doll buggy. Every year on his birthday he would get a special meal from me in an empty coffee can. He lived to be 14 years old.

I remember when I wasn't feeling good Mama would fix toast with milk and sugar on top. It did help the tummy.

I remember when polio was spreading in our area. Our neighbor Joy Wilkans contracted the disease and spent time in the children's hospital at University of Iowa in Iowa City in an iron lung. She did get well, but had several health problems into adult life. Her sister, Naomi, was my age and we as neighbors and classmates would spend lots of time together. Thankfully, no one else got polio.

I remember in the evenings after supper Mom would be crocheting or embroidering while watching TV. She did a beautiful job on both. After I joined 4H and took things to the fair, we convinced her to take some of the pillowcases she crocheted to the fair where she received several ribbons.

I remember Mom and Dad were always willing to help others. Dad made sure Aunt Myrtle had a place to live and medical help when needed. When Uncle Lloyd was living in the county home in Mason City he would leave and hitchhike to our place and Dad would make sure he got back to the county home. When a friend was having marital problems Mom and Dad gave their 3-year-old daughter, Deanna a place to live. I'm not sure how long she lived with us.

I remember every year we would get together with Aunt LaVonne and Uncle Pete to celebrate the double wedding anniversaries of Mom and Dad's and Aunt LaVonne and Uncle Pete's wedding. Aunt LaVonne and Uncle Pete lived in Minnesota so every other year was spent in Minnesota. Needless to say getting married in March we never knew what the weather would be like. I remember one year we drove home from Minnesota in a blizzard and Daddy drove with his head out the window to see the road.

In 1959 I, Mardelle, my brothers Clarence and Melvin and sister-in-law Jeanette decided to have an open house at our home for the 25th anniversary of Mom and Dad and Aunt LaVonne and Uncle Pete. Our plans were to have

25ᵗʰ wedding anniversary with Pete, LaVonne, Edith, Ralph.

a mock wedding for the couple. The day we had set for the open house there was a blizzard so that day we had to cancel and have it the next week.

I remember when I was dating there were a few times that the yard light, turned on by a switch that was in Mom and Dad's bedroom, would come on so that was the time to tell my date it was time for him to leave and I come in the house.

I remember after coming home from school I would watch Dick Clark's American Bandstand show on TV.

I remember at Easter we would go shopping and I would get a new dress and Mama would get a hat to wear to church on Easter morning.

I remember Christmas being an exciting time of the year. Christmas eve we would go to church. When we got home we would hang one of our socks on the davenport and one of our shoes below the sock to be filled by Santa. We would leave cookies and milk for Santa. Our Christmas tree was on the porch with a door with a window. The door was always closed on Christmas morning and we weren't to go in until we opened presents. Christmas day was spent at Grandma and Grandpa Hills so we had to wait until night to open gifts. It was hard waiting, looking through the window on the door,

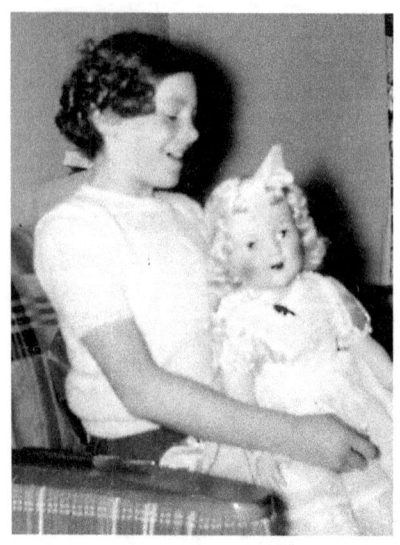

Christmas doll.

wondering if I got what I wanted. I was happy when I got a doll every Christmas. From Santa my sock would have a candy cane and a huge apple and orange in my shoe. I remember when Melvin got the bicycle he wanted. The first time he rode it he slipped on the ice and had to have stiches in his knee.

I remember going on vacation to New York City and we stayed in New Jersey. Mom was fixing breakfast with eggs we had bought in New Jersey and they were rotten. Mom was not happy. I don't remember what we had to eat then. In New York City we went on the subway train and I was so afraid the doors would close before we all got on.

I remember going on vacation to Yellowstone Park and we stayed in a cabin. At night you could hear the bears getting into the trash cans and on the roof of the cabin. There were black bears everywhere. When Dad opened the car trunk a black bear cub grabbed my little suitcase with the plastic handle. Somehow we got it back and it had the bear's teeth marks in the handle. I didn't like the smell of the mud pots. I remember Melvin ordering a peanut butter sandwich at the lodge and it came with mustard and other condiments on it. He was not happy and didn't eat it.

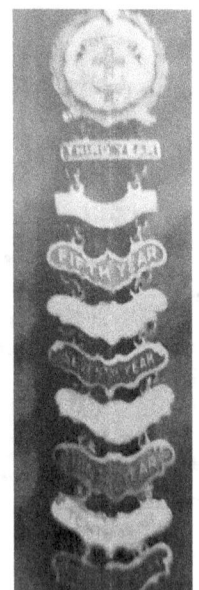

Perfect Sunday School attendance – 11 years.

I remember going on vacation to Montana. We visited Dad's friend Reinhard that he met while homesteading in Montana. The mosquitoes were so big and so many. Reinhard's two little boys wore coveralls to protect from bites and the weather was so hot.

I remember while on vacations if we stayed in a town on Sunday we went to church. I went to Sunday school so I would still have perfect attendance and get a year bar on my Methodist Church brooch pin at home.

CHAPTER 12

MEMORIES OF AN 87 YEAR OLD

(One of the Authors)

I remember staying overnight with Grandma and Grandpa Hill. We had corn flakes for breakfast. Grandpa's milk tasted different than my dad's milk.

I remember a family gathering at Grandpa Fitz's house. All the aunts and uncles were there. Uncle Hiram asked me to come closer so he could look in my ear. He reached up to my ear and magically removed a 50-cent piece from my ear, which he handed to me. I guess I looked puzzled so he did it again, with the same result. I rushed to show my coins to my mother but was told I had to give them back because they weren't mine. But Uncle Hiram said no, he had found them in Clarence's ear.

Clarence on the stump.

I remember an early morning phone call. Uncle Arthur's barn was on fire. Into the car we piled to go help.

I remember when my cousin got into the cupboard under the sink, found the Lewis Lye and decided to eat some. I remember my mother spending many afternoons sitting with her, rubbing her feet, while she recovered.

I remember when Valere lived with us for a time while her mother recovered from an illness.

I remember when Deanna lived with us for a time while her mother dealt with some personal issues.

I remember when Grandpa and Grandma Hill had a car accident. For some reason they went in the ditch and Grandpa mistakenly stepped on the accelerator instead of the brake, sending them up and over a driveway, and causing Grandma to have a whiplash injury. I remember that they lived with us for a time while Grandma recovered enough that they could manage at home. I remember that the last day I could hear Grandpa singing in the living room. "We're going home tomorrow, We're going home tomorrow," he sang to Grandma, and probably himself.

I remember a Fitz gathering when a phone call came that Cousin Burton had tipped his car over in the ditch on the way home.

I remember when cousin Janice became romantically involved with a black boy. That was before mixed-race marriages were common and it caused great anxiety for her father, my uncle. As a result she and her father became estranged, at least at the time, and that caused my mother great torment that her own brother would do that to his daughter. When my cousin and her husband or future husband moved I think to Washington or Oregon, my mother tried to stay in touch by letter to let her know that she was loved.

I remember when cousin Kathleen and her boyfriend sat in the driveway of my aunt and uncles house, doing what young people do, with the car running for heat. In the morning they were found, both unresponsive due to carbon monoxide poisoning. They both recovered but it caused much anxiety for a while.

I remember that cousin Laverne worked in a TV dinner factory, when TV dinners were a new item, and he was telling us that they had to count the number of peas that they put in each tray. Probably an exaggeration but the point was well taken. Laverne was a bit overweight and rode a motorcycle.

I remember as a young boy going with the family west on Highway 3, probably on our way to Rowan or Dows, and came upon a motorcycle accident. The rider was laying on the highway, obviously with serious injuries and great pain. That made a big impression on me and I never wanted to try riding a motorcycle and only once remember riding a motorized scooter.

I remember one evening that my mother was nagging me to get my homework done and I was using the excuse that I couldn't do it. Finally she agreed to help me as soon as the dishes were washed. So we sat down in the living room ready to go to work, I reached in my bag and the book with the assignment was not there. I guess I had left it at school. I can't quite describe how that made me feel because I really thought I had the book, but it wasn't a pleasant feeling, a feeling that has stuck with me all these years. I guess I was just so sad and upset that I had disappointed my mother.

I remember when my dad, Ralph, had taken a bucket of boiling water to the basement of the new house, set it on the floor and backed up against it splashing boiling water into his boot. Obviously a painful mistake.

I can't remember my Mom or Dad ever giving me a spanking. Maybe it is just selective memory because what I do remember is that mother could just threaten to get the stick, which was kept on a shelf above the stove, to get my attention and stop my bad behavior.

I remember that mother canned quart after quart of tomato juice and we could usually find a quart or two in the water tank in the well house to keep it cool. Oh how good that tasted on a hot afternoon.

I remember that Dad didn't use the word "zero." Instead, it was "ought." So if he was saying 1802, he would say "eighteen ought two."

I remember that Dad always put sugar on sliced tomatoes.

Cousins Alvena and Sydney

We think Cousin Alvena and Sydney Early and their family deserve mention in this book because of their residence on the farm both before and after it became the Ralph Fitz farm, and because Sydney, like Ralph, had made the difficult decision to reject a deferment in favor of serving their country. Uncle Charley, we think, rented and farmed the southeast quarter from about 1913 until 1934 when Ralph bought it. In 1915 Uncle Charley married Aunt Ethel and they lived together there until Aunt Ethel died in 1933. Alvena was born in 1921 and while we don't know if she was adopted as an infant, we do know that she was adopted by Uncle Charley and Aunt Ethel. So she, along with her sister Mary Catherine, would have spent at least some of their childhood years on that farm.

Quoting from an article written by Editor LaDona Roelfs in the *Ackley World Journal*,

> Barely out of his teens, Sydney Early of Dumont was drafted from Butler County to serve in the Army Air Corps. He could have gotten out of it …he was the only one at home to help his dad farm the land. Dad wanted to get him a farm deferment. Young Early didn't want that. He was watching friends and neighbors enlisting or being drafted, and really thought he should be a part of the group, so they asked for a month's deferment to have a farm sale. When the draft board didn't call back after the month was up, Early checked into it, saying if it was going to be a couple weeks yet, he wanted to get married. In December he and Alvena Fitz were married and Syd left for the service in January 1944.

Sydney became a gunner on a 10-man B-17 combat crew in the 775th Bombardment Squadron which flew almost every day.

> The day we were hit we were flying over an oil field in Munich where the Germans had lots of refineries. Everybody knew about that place. It was not fun. The thought of having to fly over these oil fields actually made some guys throw up in fear.

Early explained there were many enemy planes out to defend the oil fields, and his plane was being hit by a lot of flak, knocking out two engines. The crew parachuted.

> We had been hit. It took the inside engine out and the end one was screwed up. We went down two engines and too far from home to get back. We didn't have a lot of choice. The pilot and I were the last ones out, and as we jumped, it blew up.

Hitting the ground in one of those parachutes is equivalent to jumping off a 20-foot building. I landed on the top of a mountain in a forest. I found a tiny opening in the trees. Sure I was scared. We were used to being scared. We were strung out over a distance because our plane was going about 150 miles an hour. Some were able to get together, but I was alone.

Early had a map and headed to Switzerland. He almost made it, but exhaustion got the best of him, and when he woke up from his nap in a barn, a civilian he had passed on the road and two German guards were standing over him.

I ended up in a German soldiers' mountaineer training camp. I was the only one there. They threw me in a cell and later loaded me up with two guards and an interpreter on a train to Munich. I didn't know enough sense to keep my mouth shut so they beat me up pretty good.

The destination, an interrogation center. They threw me in solitary for five or six days. They knew all about our planes, but they quizzed us about them. The idea with interrogation is to break you, mind control. They then loaded everybody onto a train and went to the town (Grosstychow), three miles from the prison camp.

When we went to this Grosstychow ... that was when we ran into trouble ... going from there to the camp (Stalag Lutz IV, Poland), about three miles out. They (German soldiers) were a little upset (he laughs, sarcastically) so they fixed bayonets and had their dogs, German shepherds, trained to kill, on us. They had people stationed along the road with machine guns and started us out running to the camp. We were already worn out from being shot down and crippled up. We were forced to run and the dogs were at us.

There was one guy got into camp and he had 60 holes stuck into him. They didn't try to kill him, they just stuck him. He didn't die, but he couldn't lay down, he couldn't stand up. I arrived in camp in rather bad shape. But I discovered if you stayed in the middle of the pack, you stood a better chance. It wasn't just being poked, they'd kick you and whack you with the butt of their gun, or the dogs would bite you. You were scared. We were young.

The camp had two 10 feet high barbed wire fences and a 4 feet high rolled barbed wire barrier all surrounding the camp.

The buildings were up on stilts so they could look under them. The barracks were planned originally for 10-15 people but had 25-30 in each room. Beds were stacked two and three high, with straw mattresses. Sometimes they slept two on a mattress. Some of us slept on the floor. I slept on the floor, no mattress.

Your food was very limited and so your ambition wasn't much. We got two meals a day of watery stew of potatoes and a little meat, probably horse meat, and black bread.

What did I do all day? I thought of mean things I could do to retaliate. Of course, I couldn't do anything. It was just something I could think about.

On February 6, 6000 prisoners were divided into groups of 250 and were headed out on foot. "It started to mist, then turned to rain, then to snow. We walked out the first night, 20 miles."

The men were told it would only take three days — it ended up taking 86.

During the day we marched and at night sometimes we ended up in like a farm grove, no buildings. We each had a blanket. We moved in pairs with two men together. I got in with a guy from Alabama and we put a blanket on the ground and one on top of us. I had taken my shoes off and he had done the same.

In the morning, he had to put my shoes on me and I put his on him because we couldn't bend over, too stiff, and our feet were frozen and swelled up. We yanked the frozen blanket off the top and the one frozen to the mud beneath … and then it got worse.

Early's knee, injured when he bailed out, gave out and he and a friend tore up an old shirt and every morning they would tightly wrap his knee. He used a walking stick to hobble along. The Germans would put their dogs on the stragglers who fell behind. Finally he was forced to get on one of the "sick" wagons with other sick POWs.

Troops often marched all day with little or no midday food, water or rest. Adding to the misery was one of Germany's coldest winters ever. Snow piled knee-deep at times and temperatures plunged well below zero. By this time, the men were starved, down to skin and bones from lack of nutritious food — if they were lucky, two potatoes a day, maybe a bit of kohl-rabi and sometimes a bit of black bread. Early was down from 190 pounds, when he was drafted, to 120 pounds.

> We walked almost 600 miles until the 2nd of May, 1945. Towards the end, if we could get five miles a day we were doing good.

Of the 6000 men who started on the march, approximately 1,500 perished from disease, starvation or at the hand of German guards while attempting to escape.

The POWs thought the march would never end. 'They kept moving us and kept us marching just to keep us captive,' Early stated: 'They didn't want us liberated. They started marching us back to the camp. We were finally freed May 2nd and the war was over May 8th.'

They were flown to France where their stomachs gradually learned to hold food again. 'Our stomachs had shrunk so small you couldn't hold a decent meal. If you eat too much it swells up and can burst.' After sending them back to Patrick Henry in Virginia the war-weary men were finally put on a train to Jefferson Barracks in Missouri and headed home."

Sometime after Sydney got home our family went to visit them at their apartment which I think was in Dows. I, Clarence, remember overhearing Alvena tell my mother that Sydney was so starved for sweets that in the evening he would fill a bowl with sugar, add some cocoa, and sit and eat it.

Dad needed farm help and Sydney needed a job and so before long Sydney, Alvena and the family were living in the tenant house on our farm and Sydney was working for Dad on the farm.

It was while Sydney and I were walking beans, which means we were walking each row and pulling volunteer corn plants or large weeds that would interfere with combining, that Sydney would talk a bit about being a prisoner of war. There may have been other stories but the one that stuck in my young mind was when he told me about having to clean the latrine, which was simply a deep ditch where the German soldiers would defecate and urinate. And while the prisoners were in the ditch working, the guards would take great pleasure in urinating on them.

Quite a contrast to how the German POW that helped us pick sweet corn was treated!

CHAPTER 13

THE AMERICAN DREAM ACCOMPLISHED

Farmstead at retirement.

A Good Life, Well Lived

Ralph and Edith's family. Back row: Michael, Roger, Brian, Charles, Melvin, Allan, Jeanette, Debbie, Clarence. Front row: Mark, Mardelle, Edith, Ralph, Donna, Jonathan, Karna, Joni.

LATIMER — Funeral services for Ralph G. Fitz, 82, of Latimer were held on Friday at the United Methodist Church in Rowan with Pastor Franklin Schwarm officiating. Military burial was by the Hansen—Have Post no. 658, American Legion, Latimer, Iowa at the Hampton Cemetery. The Willim Funeral Home of Latimer was in charge of the arrangements.

Mr. Fitz was born on June 5, 1894, in Green County, Iowa and was the son of Alvin and Addie Fitz. He died May 4, 1977, at St. Joseph's Mercy Hospital in Mason City where he had been a patient for the past 4½ weeks.

He was raised at Rockwell City and then served with the U. S. Navy in World War I. After the service he homesteaded in Montana and then settled on a farm near Alexander in 1920. On March 14, 1934, he was married to Edith Hill at the home of her parents near Dows. They farmed on Highway 3 until retiring to Latimer in 1966. For many years he was a member of the United Methodist Church in Rowan. He was also a member of the Latimer American Legion and the World War I Barracks in Hamp on.

Mr. Fitz is survived by his wife, and three children, Clarence of Eagan, Minn., Melvin of Alexander and Mardelle (Mrs. Roger Meyer)of Iowa Falls; nine grandchildren, three brothers and three sisters; Ray, of Dows, Earl, Hayfield, Minn., Lloyd, Mason City, Myrtle, Hampton, Mrs. Velma Heideman, Rockwell City and Mrs. Ralph (Hazel) Marken of Waterloo. He was preceded in death by his parents, one brother and two sisters.

Ralph's obituary.

Rites held at Latimer for Mrs. Edith Fitz

LATIMER — Services for Mrs. Edith A. Fitz, 76, were held Tuesday from Rowan United Methodist Church with the Rev. David H. Wagner officiating. Burial was in the Hampton Cemetery and Willim Funeral Home of Latimer was in charge of arrangements. Mrs. Fitz died at her home Saturday.

Edith A. Fitz was born Aug. 28, 1903, in Wright County, the daughter of Clarence and Mildred Bates Hill. She attended school at Dows and married Ralph Fitz March 14, 1934. They farmed west of Latimer until retiring and moving to town in 1966. Mrs. Fitz was a member of the United Methodist Church, the UMW, Senior Citizens and a charter member of the American Legion Auxiliary of Hampton.

She is survived by two sons, Melvin of Alexander and Dr. Clarence of Eagan, Minn., a daughter, Mardelle Meyer of Iowa Falls, nine grandchildren and two great grandchildren, four sisters, Veva White of Dows, Mildred Pepper of New Providence, Lavonne Morris of Ellendale, Minn., and Evelyn Hoyington of State Center, and three brothers, Clair Hill and Laurel Hill of Dows, and Arthur Hill of Iowa Falls.

Mrs. Fitz was preceded in death by her husband and two brothers.

Edith's obituary.

Ralph (Dad) & Edith's (Mom's) graves.

DESCENDANTS OF RALPH GLENN FITZ

Generation No. 1

RALPH GLENN⁵ FITZ *(Alvin Mindon⁴, Thomas³, John², Henry¹)* (b. 1894-6-5 Greene Co., IA; d. 1977-5-4 St. Joseph Mercy Hosp., Mason City, Cerro Gordo Co., IA)

Sp. EDITH ALETA HILL (b. 1903-8-28 Wright Co., IA; m. 1934-3-14; d. 1980-1-19 home, Latimer, Franklin Co., IA)

Children:

CLARENCE RALPH FITZ (b. 1935-2-18 Scott Twp., Franklin Co., IA)

MELVIN GLENN FITZ (b. 1938-2-2 Scott Twp., Franklin Co., IA; d. 1999-5-21 home, Latimer, Franklin Co., IA)

MARDELLE MARIE FITZ (b. 1941-10-8 Scott Twp., Franklin Co., IA)

Generation No. 2

CLARENCE RALPH⁶ FITZ *(Ralph Glenn⁵, Alvin Mindon⁴, Thomas³, John², Henry¹)* (b. 1935-2-18 Scott Twp., Franklin Co., IA)

Sp. DONNA LEE GINGERICH (b. 1940-3-21; m. 1962-6-9; d. 2011-8-16 Onamia Hosp., Mille Lacs Co., MN)

Children:

KARNA JEAN FITZ (b. 1967-2-10)

JONATHAN CLARENCE FITZ (b. 1967-12-6 Edina, Hennepin Co., MN)

JONI VIRGINIA FITZ (b. 1969-2-13)

NOTE: Karna and Joni are Korean sisters adopted May 1972

MELVIN GLENN^6 FITZ *(Ralph Glenn^5, Alvin Mindon^4, Thomas^3, John^2, Henry^1)* (b. 1938-2-2 Scott Twp., Franklin Co., IA; d. 1999-5-21 home, Latimer, Franklin Co., IA)

Sp. JEANETTE KAY WESSLING (b. 1939-4-8; m. 1957-2-10)

Children:

DEBRA ANN FITZ (b. 1957-8-18 Hampton, Franklin Co., IA)

BRIAN GLENN FITZ (b. 1960-1-4 Hampton, Franklin Co., IA)

ALLAN MELVIN FITZ (b. 1962-12-19 Hampton, Franklin Co., IA)

CHARLES RALPH FITZ (b. 1964-1-17 Hampton, Franklin Co., IA)

MARDELLE MARIE^6 FITZ *(Ralph Glenn^5, Alvin Mindon^4, Thomas^3, John^2, Henry^1)* (b.1941-10-8 Scott Twp., Franklin Co., IA)

Sp. ROGER WILLIAM MEYER (b. 1939-11-12; m. 1962-1-16; d. 2019-5-6 Israel Family Hospice, Ames, Story Co., IA)

Children:

MICHAEL WILLIAM MEYER (b. 1962-10-2 Hampton, Franklin Co., IA)

MARK ROGER MEYER (b. 1966-10-6 Hampton, Franklin Co., IA)

Generation No. 3

KARNA JEAN^7 FITZ *(Clarence Ralph^6, Ralph Glenn^5, Alvin Mindon^4, Thomas^3, John^2, Henry^1)* (b.1967-2-10)

Sp. (1) JASON WILLIAM WHITE (b. unkn; m. 1991-4-5-) ((div))

Sp. (2) BRUCE CHRISTIAN TRAUTMAN (b. 1958-4-12; m.1995-6-9)

Child of Karna Fitz and Bruce Trautman:

LANA CLARE TRAUTMAN (b. 2000-11-15)

JONATHAN CLARENCE[7] **FITZ** *(Clarence Ralph*[6]*, Ralph Glenn*[5]*, Alvin Mindon*[4]*, Thomas*[3]*, John*[2]*, Henry*[1]*)* (b. 1967-12-6)

JONI VIRGINIA[7] **FITZ** *(Clarence Ralph*[6]*, Ralph Glenn*[5]*, Alvin Mindon*[4]*, Thomas*[3]*, John*[2]*, Henry*[1]*)* (b. 1969-2-13)

Sp. (1) CHAD ERIC JOHNSON (b. unkn; m. 1996-1-5) ((div))

Child of Joni Fitz and Chad Johnson:

EMILY KIRAN JOHNSON (b. 2003-12-27)

Sp. (2) ZIGMUND CARL PERET (b. 1969-12-5; m. 2014-8-23)

DEBRA ANN[7] **FITZ** *(Melvin Glenn*[6]*, Ralph Glenn*[5]*, Alvin Mindon*[4]*, Thomas*[3]*, John*[2]*, Henry*[1]*)* (b. 1957-8-18)

Sp. CRAIG ALAN HILL (b. 1955-11-23; m. 1976-2-28)

Children:

SHAWN WILLIAM HILL (b. 1977-6-28)

AMANDA ANN HILL (b. 1980-12-29)

BRIAN GLENN[7] **FITZ** *(Melvin Glenn*[6]*, Ralph Glenn*[5]*, Alvin Mindon*[4]*, Thomas*[3]*, John*[2]*, Henry*[1]*)* (b. 1960-1-4)

Sp. (1) JODIE NADINE AMES (b. unkn; m. 1978-8-unkn) ((div))

Child of Brian Fitz and Jodie Ames:

BRANDON GLENN FITZ (b. 1979-1-29)

Sp. (2) ANN LEE (b. unkn; m. 1992-5-unkn) ((div))

Children of Brian Fitz and Ann Lee:

TREVOR DEAN FITZ (b. 1993-11-24)

TYLER BRIAN FITZ (b. 1995-10-16)

Sp. (3) PATRICIA LYNN MEJIA (b. 1960-6-16; m. 2000-8-26)

ALLAN MELVIN[7] FITZ *(Melvin Glenn[6], Ralph Glenn[5], Alvin Mindon[4], Thomas[3], John[2], Henry[1])* (b. 1962-12-19)

Sp. AUDREY SUE LARSON (b. 1961-11-4; m. 1982-2-20) ((div))

Children:

RYAN ALLAN FITZ (b. 1989-4-11)

JUSTIN LEE FITZ (b. 1991-10-14)

CHARLES RALPH[7] FITZ *(Melvin Glenn[6], Ralph Glenn[5], Alvin Mindon[4], Thomas[3], John[2], Henry[1])* (b. 1964-1-17)

Sp. (1) JULIE BUFFALOHEAD RHOTEN (b. unkn; m. 1988-8-22) ((div))

Child of Charles Fitz and Julie Rhoten:

LEIGHTON CHARLES FITZ (b. 1988-7-15; d. 2014-4-17 City of Iowa Falls, Hardin Co., IA)

Sp. (2) BECKY JEAN MAYLAND (b. unkn; m. 2004-5-1)

Child of Charles Fitz and Becky Mayland:

LEVI FITZ (b. unkn)

MICHAEL WILLIAM[7] MEYER *(Mardelle Marie[6] Fitz, Ralph Glenn[5], Alvin Mindon[4], Thomas[3], John[2], Henry[1])* (b. 1962-10-2)

Sp. CHERYL MICHELLE WINSLOW (b. 1973-6-23; m. 2002-9-9)

Children:

ADELINE ROSE MEYER (b. 2011-10-26)

BENJAMIN MICHAEL MEYER (b. 2016-10-27)

MARK ROGER[7] MEYER *(Mardelle Marie[6] Fitz, Ralph Glenn[5], Alvin Mindon[4], Thomas[3], John[2], Henry[1])* (b. 1966-10-6)

Sp. (1) AMY KNOCHEL (b. 1968-5-27; m. 1988-5-28; d. 2018 6-1 Pine River, Lincoln Co., WI) ((div))

Children of Mark Meyer and Amy Knochel:

NATHAN WILLIAM MEYER (b. 1992-9-1)

NICHOLAS WILLIAM MEYER (b. 1998-2-7)

Sp. (2) LAURIE ANN WILL (b.1967-9-3; m. 2001-3-9) ((div))

Child of Mark Meyer and Laurie Will:

AMANDA LEE MEYER (b. 2000-11-14)

Generation No. 4

LANA CLARE[8] TRAUTMAN *(Karna Jean[7]Fitz, Clarence Ralph[6], Ralph Glenn[5], Alvin Mindon[4], Thomas[3], John[2], Henry[1])* (b. 2000-11-15)

EMILY KIRAN[8] JOHNSON *(Joni Virginia[7] Fitz, Clarence Ralph[6], Ralph Glenn[5], Alvin Mindon[4], Thomas[3], John[2], Henry[1])* (b. 2003-12-27)

SHAWN WILLIAM[8] HILL *(Debra Ann[7]Fitz, Melvin Glenn[6], Ralph Glenn[5], Alvin Mindon[4], Thomas[3], John[2], Henry[1])* (b. 1977-6-28)

Sp. MICHELLE MARIE BURNS (b. 1978-7-19; m. 2001-6-17)

Children:

COLIN WILLIAM HILL (b. 2003-2-16)

LAUREN MARIE HILL (b. 2007-4-26)

AMANDA ANN [8] HILL *(Debra Ann[7] Fitz, Melvin Glenn[6], Ralph Glenn[5], Alvin Mindon[4], Thomas[3] John[2], Henry[1])* (b. 1980-12-29)

Sp. DANIEL BLAINE TESKE (b. 1980-9-3; m. 2002-8-17)

Children:

SHAYLA ROSE TESKE (b. 2001-7-3)

JAMES GLEN TESKE (b. 2005-8-20)

CALEB ALAN TESKE (b. 2007-6-28)

BRANDON GLENN[8] FITZ *(Brian Glenn[7], Melvin [6], Ralph Glenn[5], Alvin Mindon[4], Thomas[3], John[2], Henry[1])* (b. 1979-1-29)

TREVOR DEAN[8] FITZ *(Brian Glenn[7], Melvin Glenn[6], Ralph Glenn[5], Alvin Mindon[4], Thomas[3], John[2], Henry[1])* (b. 1993-11-24)

TYLER BRIAN[8] FITZ *(Brian Glenn[7], Melvin Glenn[6], Ralph Glenn[5], Alvin Mindon[4], Thomas[3], John[2], Henry[1])* (b. 1995-10-16)

RYAN ALLAN[8] FITZ *(Allan Melvin[7], Melvin Glenn[6], Ralph Glenn[5], Alvin Mindon[4], Thomas[3], John[2], Henry[1])* (b. 1989-4-11)

JUSTIN LEE[8] FITZ *(Allan Melvin[7], Melvin Glenn[6], Ralph Glenn[5], Alvin Mindon[4], Thomas[3], John[2], Henry[1])* (b. 1991-10-14)

LEIGHTON CHARLES[8] FITZ *(Charles Ralph[7], Melvin Glenn[6], Ralph Glenn[5], Alvin Mindon[4], Thomas[3], John[2], Henry[1])* (b. 1988-7-5; d. 2014-4-17 City of Iowa Falls, Hardin Co., IA)

Children of Leighton Fitz as per his obituary:

HAYDEN FITZ (b. unkn)

LEIGHLA FITZ (b. unkn)

DAHLIA GROSS-WILLIAMS (b. unkn)

CONNOR FITZ (b. unkn)

MALRIELLE FITZ (b. unkn)

LEVI[8] FITZ *(Charles Ralph[7], Melvin Glenn[6], Ralph Glenn[5], Alvin Mindon[4], Thomas[3], John[2], Henry[1])* (b. unkn)

ADELINE ROSE[8] MEYER *(Michael William[7], Mardelle Marie[6] Fitz, Ralph Glenn[5], Alvin Mindon[4], Thomas[3], John[2], Henry[1])* (b. 2011-10-26)

BENJAMIN MICHAEL[8] **MEYER** *(Michael William*[7]*, Mardelle Marie*[6] *Fitz, Ralph Glenn*[5]*, Alvin Mindon*[4]*, Thomas*[3]*, John*[2]*, Henry*[1]*)* (b. 2016-10-27)

NATHAN WILLIAM[8] **MEYER** *(Mark Roger*[7]*, Mardelle Marie*[6] *Fitz, Ralph Glenn*[5]*, Alvin Mindon*[4]*, Thomas*[3]*, John*[2]*, Henry*[1]*)* (b. 1992-9-1)

NICHOLAS WILLIAM[8] **MEYER** *(Mark Roger*[7]*, Mardelle Marie*[6] *Fitz, Ralph Glenn*[5]*, Alvin Mindon*[4]*, Thomas*[3]*, John*[2]*, Henry*[1]*)* (b. 1998-2-7)

AMANDA LEE[8] **MEYER** *(Mark Roger*[7]*, Mardelle Marie*[6] *Fitz, Ralph Glenn*[5]*, Alvin Mindon*[4]*, Thomas*[3]*, John*[2]*, Henry*[1]*)* (b. 2000-11-14)

Child of Amanda Meyer and Dakota Osborne-Yeager:

KYLAN REID YEAGER (b. 2021-7-13)

Generation No. 5

COLIN WILLIAM[9] **HILL** *(Shawn William*[8]*, Debra Ann*[7] *Fitz, Melvin Glenn*[6]*, Ralph Glenn*[5]*, Alvin Mindon*[4]*, Thomas*[3]*, John*[2]*, Henry*[1]*)* (b. 2003-2-16)

LAUREN MARIE[9] **HILL** *(Shawn William*[8]*, Debra Ann*[7] *Fitz, Melvin Glenn*[6]*, Ralph Glenn*[5]*, Alvin Mindon*[4]*, Thomas* [3]*, John*[2]*, Henry*[1]*)* (b. 2007-4-26)

SHAYLA ROSE[9] **TESKE** *(Amanda Ann*[8]*, Debra Ann*[7] *Fitz, Melvin Glenn*[6]*, Ralph Glenn*[5]*, Alvin Mindon*[4]*, Thomas*[3]*, John*[2]*, Henry*[1]*)* (b. 2001-7-3)

JAMES GLEN [9] **TESKE** *(Amanda Ann*[8]*, Debra Ann*[7] *Fitz*, Melvin Glenn[6], Ralph Glenn[5], Alvin Mindon[4], Thomas[3], John[2], Henry[1]) (b. 2005-8-20)

CALEB ALAN[9] **TESKE** *(Amanda Ann*[8]*, Debra Ann*[7] *Fitz, Melvin Glenn*[6]*, Ralph Glenn*[5]*, Alvin Mindon*[4]*, Thomas*[3]*, John*[2]*, Henry*[1]*)* (b. 2007-6-28)

HAYDEN[9] **FITZ** *(Leighton Charles*[8]*, Charles Ralph*[7]*, Melvin Glenn*[6]*, Ralph Glenn*[5]*, Alvin Mindon*[4]*, Thomas*[3]*, John*[2]*, Henry*[1]*)* (b. unkn)

LEIGHLA[9] **FITZ** *(Leighton Charles*[8], *Charles Ralph*[7], *Melvin Glenn*[6], *Ralph Glenn*[5], *Alvin Mindon*[4], *Thomas*[3], *John*[2], Henry*[1]*) (b. unkn)

DAHLIA[9] **GROSS-WILLIAMS** *(Leighton Charles*[8], *Charles Ralph*[7], *Melvin Glenn*[6], *Ralph Glenn*[5], *Alvin Mindon*[4], *Thomas*[3], *John*[2], Henry*[1]*) (b. unkn)

CONNOR[9] **FITZ** *(Leighton Charles*[8], *Charles Ralph*[7], *Melvin Glenn*[6], *Ralph Glenn*[5], *Alvin Mindon*[4], *Thomas*[3], *John*[2], Henry*[1]*) (b. unkn)

MALRIELLE[9] **FITZ** *(Leighton Charles*[8], *Charles Ralph*[7], *Melvin Glenn*[6], *Ralph Glenn*[5], *Alvin Mindon*[4], *Thomas*[3], *John*[2], Henry*[1]*) (b. unkn)

KYLAN REID[9] **YEAGER** *(Amanda Lee*[8], *Mark Roger*[7], *Mardelle Marie*[6] *Fitz, Ralph Glenn*[5], *Alvin Mindon*[4], *Thomas*[3], *John*[2], Henry*[1]*) (b. 2021-7-13)

CLARENCE'S FAMILY

Generation No. 2

Back row: Donna, Karna, Jonathan, Clarence. Front: Joni.

MELVIN'S FAMILY

Generation No. 2

Charles, Debra, Jeanette, Brian, Melvin and Allan.

MARDELLE'S FAMILY

Generation No. 2

Roger, Mardelle, Mark and Michael.

EPILOGUE

Writing this book has been a labor of love, first in exploring the Fitz family name in both England and America, then tracing the migration west to the Iowa destinations, then Ralph's adventures in Montana, interrupted by World War I, and finally to the Franklin County, Iowa farm where we grew up. At times it was kind of like reliving our childhoods and remembering our Mom and Dad. Working with my sister, Mardelle on this project has been rewarding and I believe has brought us closer together after our lives didn't really cross much between childhood and now.

It was fun peeking behind the words and behind the objects that Ralph, our Dad, saved and from that weaving his story as best we could with no one to ask for confirmation. Like most youngsters growing up, we didn't ask enough questions, and for that we are regretful. Ralph may have seemed at times like a simple hard-working farmer in his bib overalls, but he was sophisticated beyond his formal schooling.

He forged out on his own, probably lured by the likes of James J. Hill, the rail-

Ralph in bib overalls.

road magnate, of the promises of bountiful harvests waiting for a homesteader who was willing to work hard, into the tail end of the Montana homestead rush, only to be saved, we think, by serving his country in the navy and by the friendships he had cultivated while there.

It appears, with mostly legal documents as evidence, that there were unexplained dynamics playing out in Ralph's family, our family, probably related mostly to economics. He survived the depression years, we believe, by hard work and smart decisions. Ralph apparently took advantage of financial opportunities that arose but as we have pointed out in the book, he was always ready to help out when family members needed help.

When I was first thinking about writing this book I, Clarence, made my first trip to Montana in 2002 along with my wife Donna and another couple who were friends of ours. We visited the town of Whitewater, stopped at the only viable business, a sort of general store, and visited with the people there. I had wished later that I had taken some pictures but I didn't. We joked that when we drove into town the population of Whitewater doubled. I do remember that there was a railroad depot there that was in disrepair. That reminded me that I had heard my Dad mention that the town of Whitewater had been moved when the railroad was built to its current location so they could be on the rail line instead of on the creek. That prompted me to research the history of Whitewater and what I learned was that Whitewater first got its name from the stream that flowed through town, the Whitewater Creek. In 1927, the Great Northern Railroad built a branch line seven miles west of Whitewater. That prompted N.J. Brandt to move his store and the post office to its present location.

Whitewater Creek

Then in 2021 I made another visit to Montana and to Whitewater, having already started this book by then. The town of Whitewater had obviously lost what little luster it had during those almost twenty years. The two grain elevators, symbols of a bygone era of grain farming, were closed and in disrepair, the general store was nowhere to be found and the only building that looked like

Church in Whitewater, MT.

it might still be a viable business had a sign on the front that said "Café," but it was closed. There were two buildings still in good condition, the school and the church.

School in Whitewater, MT.

Town of Whitewater, MT, 2021.

Ralph's homestead – BLM land.

As I drove around the countryside it became very evident that my Dad, Ralph, had made the right decision in abandoning his Montana farming plans. I saw very few, sort-of, farmsteads. It was mostly grasslands like it had been before the homesteading craze. Most of the area where Ralph's homestead was located had reverted back to the United States government and was managed by the Bureau of Land Management.

In addition to the Montana trip I, Clarence, made trips to Franklin County and Greene County, Iowa gathering information and my co-author, Mardelle, made similar trips to Franklin County and Calhoun County, Iowa. What we observed as we traveled was that the contrast was stark between the depressed appearance of the farming communities around Whitewater, Montana and prosperous appearance of those in Franklin, Greene and Calhoun Counties in Iowa. Our Dad, Ralph, had made the right decision.

If you have gotten this far in the book you already know that there were some pages of boring reading as we described legally certain parcels of land, and it's okay that you skipped over those paragraphs. But for anyone who wants to pinpoint certain areas, you now have the legal data without having to redo the research.

We do hope that someone at some point will pick up the baton and pursue the unanswered question of whether our family actually originated from the FitzRandolph's and became the Fitz's by dropping the Randolph as my co-author Mardelle and I suspect.

We have enjoyed writing this book. We trust you have enjoyed reading it.

Flag that draped Ralph's casket.

BIBLIOGRAPHY

1. *Fitts Families (Fitts-Fitz-Fittz) A Genealogy,* by Sylvia Fitts Getchell – 1989

2. *"…and the Mille Lacs who have no Reservation…"* by Clarence Ralph Fitz – 2016

3. *Love and Hate in Jamestown* by David Price - 2003

4. *Genealogy of the Fitts or Fitz Family in America* by James Hill Fitts – 1869

5. *Fitz Family History and Genealogy* by Carrie Fitz Lambert – Unpublished

6. *Dead Wake* by Erik Larson – 2015

7. *The First World War* by Jere Clemens King – 1972

8. *Jane's Fighting Ships of World War I* by Fred T. Jane – 1914

9. *The History of World War I Naval Warfare 1914-1918* by Tim Benbow – 2008

10. *The Experience of World War I* by J.M. Winter – 1989

11. *Campfires and Cowchips* by Floyd Hardin – 1972

12. *Montana Memoir* by William L and Sandra V. McGee – 2016

13. *History of Franklin County, Iowa, a Record of Settlement, Organization, Progress and Achievement* by I.L. Stuart – 1914

14. *The Doughboys* by Gary Mead – 2000

15. *The Prison Camp at Andersonville* by Stephanie Steinhorst & Chris Barr - 2014

OTHER BOOKS BY THIS AUTHOR

"...And the MILLE LACS who have no reservation..." — *A history of the Chippewa Indians in Mille Lacs County, Minnesota up to 1934,* Fideli Publishing Inc., 2016

The PENDULUM...from Indian Removal to buying Mille Lacs, Fideli Publishing Inc., 202